The Practical Application of the Process Capability Study

Evolving from Product Control to Process Control

The Practical Application of the Process Capability Study

Evolving from Product Control to Process Control

Douglas B. Relyea

CRC Press is an imprint of the
Taylor & Francis Group, an **informa** business
A PRODUCTIVITY PRESS BOOK

Productivity Press
Taylor & Francis Group
270 Madison Avenue
New York, NY 10016

Printed in the United States of America on acid-free paper
10 9 8 7 6 5 4 3 2 1

International Standard Book Number: 978-1-4398-4778-7 (Paperback)

Visit the Taylor & Francis Web site at
http://www.taylorandfrancis.com

and the Productivity Press Web site at
http://www.productivitypress.com

Contents

Introduction

The *process capability study* is a powerful tool that, when understood, will provide benefits to almost every department within a manufacturing organization.

A wealth of information is available on the technical concepts of the process capability study. Much of what is available has placed a great deal of emphasis on the mathematics and technical nuances that would be of interest to quality professionals, engineers, and statisticians. Unfortunately, concentrating on the math and fine distinctions, such as the difference between alpha- and beta-type errors, has created barriers preventing many people from fully appreciating the basic concepts, the simplicity, and the usefulness of the tool.

One significant casualty of the narrow acceptance and use of the process capability study has been an appropriate return on investment many companies have failed to realize from their statistical process control (SPC)/Six Sigma effort. In many companies, the SPC/Six Sigma effort has evolved into bulletin boards filled with bar graphs, Pareto charts, and project team reports addressing nonmanufacturing-related concerns. Using well-defined group dynamics and problem-solving activity to address organizational concerns can provide significant benefits to a company. However, the essence of the original SPC/Six Sigma approach was directed at understanding and reducing manufacturing process variation.

Through the SPC/Six Sigma philosophy of reducing variation in the manufacturing process, the biggest return on investment will be realized. Reduced raw material consumption and increased productivity are only two of the benefits of a reduction in process variation.

One proven method of reducing process variation is the application of the SPC technique known as the process capability study. However, this important tool is not widely or properly used in much of American manufacturing today. I believe the primary reason for the process capability study falling into disuse is the lack of understanding, outside of quality assurance departments, of the basic concepts and benefits of the tool.

I wrote this book with specific readers in mind. Those readers would be engineers and members of upper and middle management, production supervisors, and operators who are involved in making multiple manufacturing decisions on a daily basis. I also wrote this book for the purchasing manager who is responsible for the acquisition of quality raw material at the lowest cost. And I had in mind the sales and marketing professionals who would benefit from increased sales if they had the ability to promote higher-quality product at a lower cost through the presentation of factual product data such as a capability index (Cpk). I wrote this book for every professional who wants to make his or her company more competitive but does not need to understand the complex mathematics that goes along with many aspects of SPC/Six Sigma.

My intent is to offer concepts without the mathematics that accompany them. I rely on some simple arithmetic in some instances but nothing more complicated than dividing one number by another.

Where appropriate, I indicate that more detailed information regarding certain concepts may be found in the Appendix. I recognize that some readers may want to understand more about the mathematics, but I did not clutter up the body of the text with Greek letter alphabet soup and mathematical symbols.

Also, for those interested readers who wish to perform a process capability study or a measurement process analysis, two Excel® spreadsheets are provided online (available at www.dougrelyea.com) to facilitate their endeavors. Once the appropriate data are entered on the spreadsheet, all of the math is performed, and a bottom-line conclusion is presented, in easy to understand language, without having to go to sophisticated statistical software.

In all examples concerning dimensional studies, the units are in inches.

My fervent hope is that I have presented basic concepts without being too basic.

Chapter 1

Review of Basic Concepts

There are people who are afraid of clarity, because they fear it may not seem profound.

Elton Trueblood

Variation and Specifications

In the manufacturing world, productivity, quality, competitiveness, and customer satisfaction are all determined by the degree of variation found in finished product.

As a practical matter, all manufacturers are customers as well as suppliers. Manufacturers process purchased raw material, and they provide customers with finished product. As a result of this dual role, manufacturing companies need to be acutely aware of the variation to be found in purchased raw material as well as the variation inherent in their finished product.

As buyers of components and raw materials, manufacturers have always understood the importance of imposing some control over the variation in the critical characteristics of their purchases. For much of our manufacturing history, simple blueprint specifications and industry standards satisfied the need for this control. These specifications and industry standards identified the most desired values of tensile strength, density, melt index, diameter, and so forth, deemed to be necessary to the optimum function of the end product. This most desired value became known as the nominal specification.

Apparently, everyone intuitively accepted the fact that nothing can be made without some degree of variation, so design engineers and purchasing agents provided some latitude to the nominal specification in the form of upper and lower specifications.

So, the nominal specification represents the value most desired by the customer, and the upper and lower specifications represent the worst case of what the customer is willing to accept.

Two worrisome aspects of customer-supplied specifications seem to be virtually universal.

First, in spite of the nominal specification being recognized as the value most desired by the customer, many companies feel quite comfortable providing product that consumes the entire range of upper to lower specification—sometimes within the same shipment.

Second, many manufacturing facilities regularly agree to provide a customer with product that must fall within the customer's specification range without appropriate data to validate the commitment.

As simple as these two concerns appear to be, they form the basis for the need to understand and employ process capability studies.

Product Control versus Process Control

Today most people in industry understand that product variation is a direct result of process variation, and yet efforts to control the product seem to far outweigh efforts to control the process.

Understanding the effects that process variation has on critical product characteristics is the initial goal of a process capability study. Specifically, a process capability study is performed in order to determine if the *process* variation is compatible with producing consistent *product* variation (stability) and is capable of maintaining the consistent *product* variation so that it falls well within customer specification (capability). A statistical process control (SPC) chart, created as a result of a process capability study, requires product data to be plotted in relation to control chart elements such as the centerline, upper control limit, and lower control limit. Certain patterns of product data relative to the three control chart elements (centerline, upper limit, and lower limit) indicate if the process is stable or unstable. And the application of simple arithmetic to product data resulting from the process capability study informs us if the process is capable.

If the initial control chart indicates the process is unstable, the first order of business is to identify and remove the cause of the instability. This requires an investigation to identify and correct the process parameter (line speed, pressure settings, pH level, etc.) causing the product instability. Once the process is determined to be stable, the capability of the process can then be arithmetically determined. Going forward, the control chart can then be used to alert operators and supervisors to incidents of process instability.

In essence, the SPC chart that results from a process capability study enables us to analyze product variation for the purpose of understanding process variation. From this, we could conclude the ultimate purpose of the process capability study, and the ensuing control charts, is to understand and eventually control key process parameters.

As previously stated, certain patterns formed by *product* data recorded on a shop floor control chart serve the purpose of alerting the operators to incidents of *process* instability. These incidents of *process* instability, if investigated, will lead to an understanding of the *process* parameter causing the *product* instability. Once a key *process* parameter has been identified, a control chart can then be created for that *process* parameter; this would be a true statistical process control chart.

Today, in manufacturing, responses of operators and supervisors to control chart indications of instability generally result in *process* adjustments to compensate for indications of *product* instability. Short-term concern for product control seems to outweigh the long-term benefit that would be derived from identifying key process parameters and evolving from product control to process control.

If control chart indications of instability result only in adjustments to a process parameter and do not result in an investigation of why a process parameter is causing the product instability, then the ultimate goal of process capability studies and SPC is being undermined. The nomenclature *statistical process control* implies that efforts should be directed at controlling the process, not perpetually attempting to control the product by means of compensatory adjustments to the process.

If the sole purpose of a SPC chart is to alert the operators to make process adjustments, the *process* control chart is not serving the purpose for which it is intended.

What many companies today refer to as SPC charts are, in fact, "statistical product control" charts. A statistical product control chart is characterized by the fact that operators and supervisors make adjustments to process parameters only when they see indications of instability on the control chart and make no effort to record the process parameters that were adjusted or to otherwise attempt to identify the root cause of the product instability. For example, if the process parameter line speed is adjusted every time the control chart indicates the product characteristic length has become unstable, and no attempt is made to identify and correct the root cause—perhaps a faulty motor controller—statistical *product* control is at work.

For purposes of clarity throughout this book, I will refer to those charts that are monitoring product characteristics as *statistical product control charts*, and I will refer to those charts that are monitoring process parameters as *statistical process control charts.*

Going forward, please remember the following:

A *statistical product control* chart is a graphic representation of the variation present in a product characteristic resulting from the measurement of random samples selected during the manufacturing process.

A *statistical process control* chart is a graphic representation of the variation of a specific process parameter that is known to cause instability or incapability of an important product characteristic.

The number of statistical product control charts and the lack of statistical process control charts on manufacturing shop floors is evidence of the preceding statement that too many companies concentrate on product control instead of process control.

The concentration on product control is what has led many organizations down the path of never-ending and expensive SPC chart bureaucracies. Numerous manufacturing facilities take pride in pointing out to a visiting customer multiple "statistical process control" charts attached to certain pieces of equipment. Usually each of these posted charts, maintained by one very busy operator, would represent an attempt to understand and control a single *product* characteristic important to the customer. When the production line is changed over to a different product, the active charts are filed, and new charts with preprinted upper and lower control limits developed for the product about to be produced are posted. Then data for the product currently being produced are plotted on the newly posted charts.

When the statistical product control chart indicates unusual variation in the product characteristic being monitored, the operators and supervisors, as instructed, will compensate by making adjustments to the process. Usually, no effort is made to identify the process parameter that resulted in the change in product variation.

Somewhere along the way, we lost sight of the goal, which is *process control* not *product control.* And process control begins with a process capability study.

The ultimate goal of the process capability study concept is rather simple. The variation of a key product characteristic is studied, resulting in a statistical product control chart. The data patterns on the statistical product control chart indicate if the process was stable or unstable during the time the study was conducted. If the process is stable, simple arithmetic is employed to determine if the process is capable or incapable.

When the process capability study indicates the product is unstable or incapable, steps must be taken to identify the process parameter causing the product instability or incapability.

Once the parameter causing the instability or incapability is identified and corrected, SPC charts are established to understand, control, and improve the key process parameter contributing to product variation.

Benefits of Process Capability Studies

Regular use of process capability studies provides the following benefits to the shop floor:

- The elimination of numerous product control charts in favor of a few process control charts.
- Process knowledge will be acquired, which will facilitate problem solving and introduction of new product.
- It provides a means to easily qualify new raw material, which may or may not increase product variation. When a new raw material is being considered, a process capability study can be performed, and results using the new material may be compared to results with existing material.
- Shop floor associates will better appreciate that process parameters normally vary, which tends to reduce or eliminate unnecessary process adjustments.
- Some new product lines may be quickly determined to be unstable or not capable and not yet ready for production.

Other process capability study benefits to manufacturing, engineering, sales, marketing, and purchasing will be discussed in later chapters.

Industry would be well advised to concentrate on process control in favor of process control. And, of course, the first step to process control is to perform a process capability study.

Topics Covered Going Forward

The following will be presented in this book:

- Detailed discussions of the concepts briefly touched on in this chapter
- Basics of statistical process control
- A step-by-step procedure to properly perform and analyze the results of a variable process capability study
- A step-by-step procedure to properly perform and analyze the results of an attribute process capability study
- How to qualify a measurement system
- How to identify the causes of instability and incapability
- How to use the process capability study to solve shop floor problems
- Case studies from a number of different technologies, including extrusion, molding, automatic assembly, cabling, machining, manual assembly, and fiber optic manufacturing
- How to use Excel to determine if a process modification has truly improved the output
- How to apply the process capability study concept to sales, marketing, and purchasing

Chapter 2

Short Course in Variation

Fast is fine but accuracy is everything.

Wyatt Earp

Stability and Normal Variation

Among the benefits of performing process capability studies is the unique feature that the studies will provide the answers to several important questions. And the answers are such that they can bridge many gaps that may exist between the shop floor and other facets of an organization regarding the manufacturing processes.

With respect to critical product characteristics being studied, the specific questions that can be factually answered are as follows:

- Does the process demonstrate stability?
- Does the process demonstrate capability?
- Can we describe capability with a number?

Stability

The dictionary defines *stability* as "1. Firmness in position. 2. Continuance without change; permanence." (http://dictionary.com)

Many people relate best to the term *stability* when it is presented in terms of histogram data demonstrating a pattern of normal variation.

Normal

The dictionary defines *normal* as "1. Conforming to the standard or the common type; usual; not abnormal; regular; natural. 2. Serving to establish a standard." (http://dictionary.com)

Normal Variation

Normal variation is impossible to avoid. If we were to construct histograms for the types of variation we experience every day, we would discover that most of the histograms would demonstrate normal variation, or, as many would say, the data would be *normally distributed.*

For example, most people would discover the time it takes them to get to work each day over a one-year period would be normally distributed. The heights of high school seniors would prove to be normally distributed, as well. Imagine attending a graduation and every young man in the graduating class was exactly 5 foot, 8 inches tall? This would be a virtual impossibility. If we were to measure the heights of a class of 54 young men graduating from high school, and round off to the nearest inch, we would notice that the average height would be about 5 foot, 8 inches, a few of the basketball players would log in at about 6 foot, 2 inches, and there would be the place kickers and lightweight wrestlers measuring in the area of 5 foot, 2 inches. The remainder of the class would very likely fill in the two extremes in such a way as to form the normally distributed histogram illustrated by Figure 2.1.

This particular bell-shape distribution is found throughout nature and is usually termed the *normal distribution* or the *normal curve.*

There is an important convention concerning the manner in which a normal distribution is described. The center of any normal distribution is referred to as the average, and the width of the distribution is provided in terms of standard deviation or sigma. Outside the world of statisticians, the two terms *standard deviation* and *sigma* are used interchangeably. Every normal distribution has exactly three sigma (standard deviation) on either side of the center (average).

This convention of describing a normal curve in terms of sigma makes for ease of communication when discussing data. If a supplier states that his or her process is centered at 0.5 with a sigma of 0.001, it is clear that the supplier can produce product between 0.497 and 0.503.

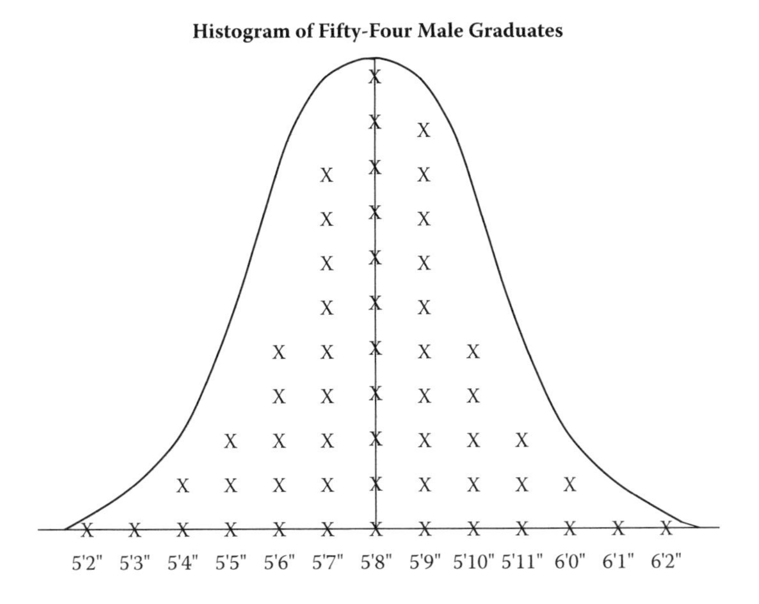

Figure 2.1 Histogram of 54 male graduates.

$$\text{Process average} = 0.5$$
$$\text{One sigma} = 0.001$$

Three sigma either side of the average

$$0.001 \times 3 = 0.003$$
$$0.5 \pm 0.003 = 0.497/0.503$$

Our manufacturing processes yield product that, when produced during a period of process stability, also demonstrate normal variation. Table 2.1 represents the diameters of 60 machined shafts, and Figure 2.2 represents the histogram of the measured diameters.

Control Chart

Although the histogram is a useful tool, we can derive more information about the process that produced the 60 machined parts if we plot the data from Table 2.1 on the Individuals product control chart illustrated by Figure 2.3.

The histogram provides information regarding only the 60 samples. The control chart, on the other hand, provides us with the upper and lower

Table 2.1 Sixty Machined Shaft Diameters

0.4998	0.4995	0.4998	0.5011	0.4990	0.4983
0.4992	0.4989	0.4998	0.4989	0.4994	0.5010
0.5002	0.5024	0.5009	0.5004	0.4990	0.4992
0.5008	0.4994	0.4999	0.5006	0.4998	0.4996
0.5008	0.4995	0.4999	0.4980	0.5000	0.5005
0.5011	0.4986	0.4997	0.4999	0.5000	0.5003
0.4986	0.4996	0.5013	0.4997	0.4998	0.5006
0.4998	0.4997	0.5006	0.5004	0.5014	0.5004
0.5007	0.5001	0.5016	0.4997	0.4988	0.4991
0.4993	0.4998	0.4996	0.5005	0.4995	0.4993

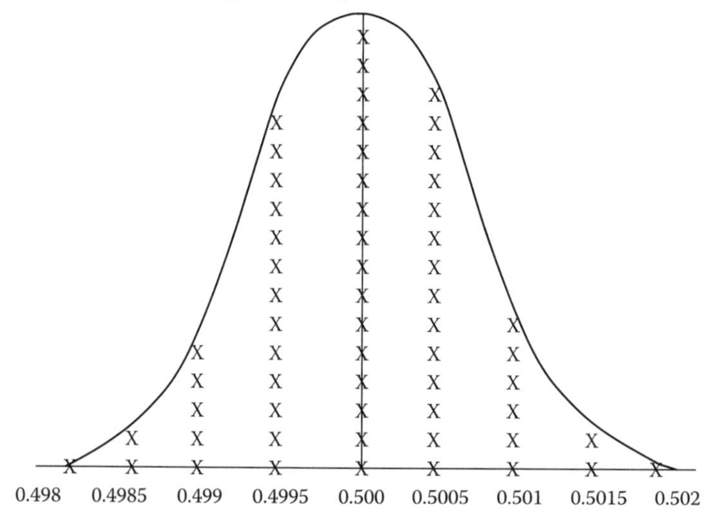

Figure 2.2 Histogram of 60 machined diameters.

control limits that predict the total amount of normal variation we would find if we measured *all* the parts made during the time the 60 samples were selected.

In this example, 60 parts were selected at random during 4 hours of a normal production run. The 60 samples were measured, and some simple

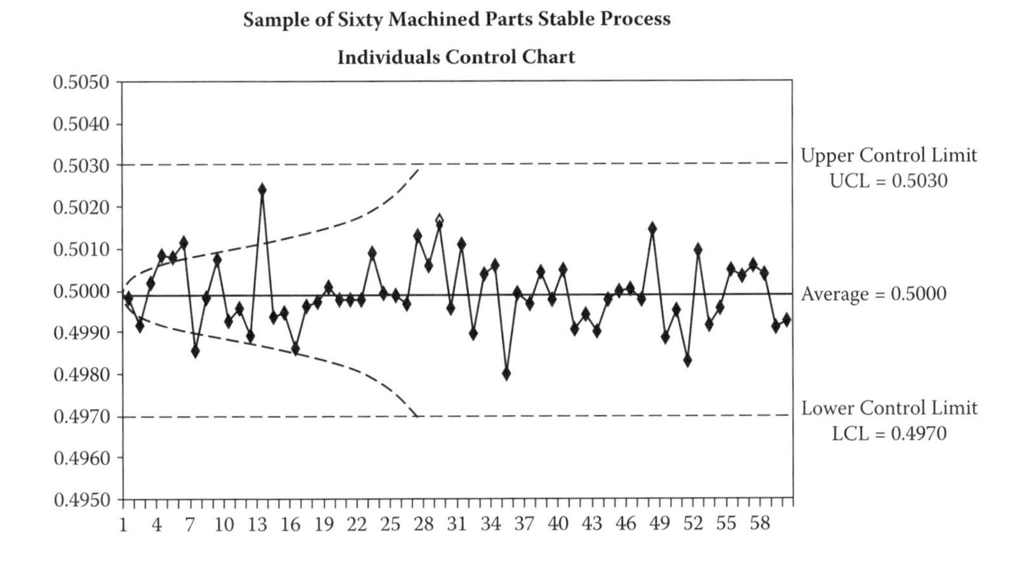

Figure 2.3 Sample of 60 machined parts stable process, individuals control chart.

arithmetic was performed, using those measurements, to determine the average diameter and the predicted upper and lower control limits. The centerline of Figure 2.3 represents the average diameter of all the parts produced, and the two dotted lines represent the upper control limit (UCL) and the lower control limit (LCL). The normal curve has been superimposed on Figure 2.3 to emphasize the fact that the upper and lower control limits, in fact, are the plus and minus three sigma limits of the normal curve. (Note that it is conventional to draw six dotted lines on the Individuals control chart to represent each increment of plus and minus three sigma. In this book, I chose not to recognize this convention so as not to clutter up the control charts.)

Because the data points on the control chart are randomly distributed about the centerline without exceeding the upper and lower control limits and abide by several other simple rules, we can say the data demonstrate a certain "firmness of position"—the process was *stable* during the time the 60 machined parts were produced.

Because the process was stable during the production run, we can use the plus and minus three sigma upper and lower control chart limits in Figure 2.3 to predict that we would find a small percentage of the total diameters produced as low as 0.497 and a small percentage as large as 0.503.

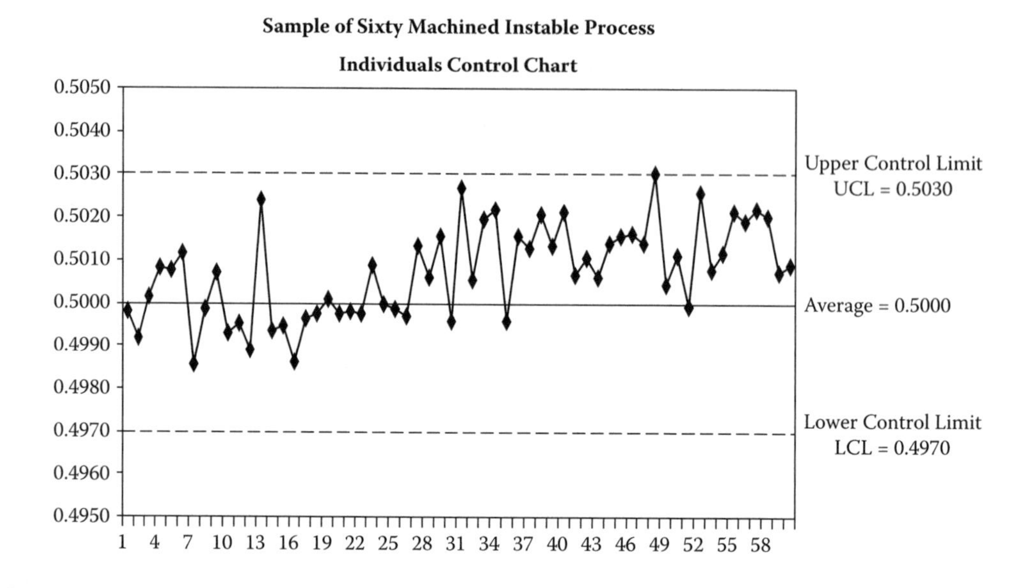

Figure 2.4 Sample of 60 machined instable process, individuals control chart.

The arithmetic necessary to calculate the control chart average, upper, and lower limits for this Individuals control chart may be found in the Appendix "Chapter 2, Individuals control chart calculation."

Instability

The dictionary defines *instability* as "1. The quality or state of being unstable; lack of stability or firmness. 2. The tendency to behave in an unpredictable, changeable, or erratic manner." (http://dictionary.com)

Figure 2.4 represents the same product characteristic, but for discussion purposes we will consider this process as having experienced instability shortly before the lunch break.

Note that about data point number 25, the measured diameters shift upward, and they are no longer randomly distributed about the centerline of the control chart. A shift such as this is an indication of instability.

When the statistical product control chart indicates instability, it is a result of one or more of the process parameters becoming unstable, erratic, or unpredictable. Perhaps the line speed is changing erratically or the temperature controllers are acting in an unpredictable manner. When the process parameters are behaving in an inconsistent, unpredictable manner, we cannot have any confidence regarding the output of the process. Due to the instability, we can make no predictions regarding the largest or smallest diameter produced during the period of instability.

Process instability is a problem. Anytime we recognize process instability as a result of the statistical product control chart, we have a responsibility to identify the process parameter that is behaving erratically and correct it.

As mentioned in Chapter 1, the basic concept of statistical process control is to utilize statistical product control charts to identify periods of process instability. Then identify the instable process parameter, improve the process parameter, and eventually replace the statistical product control chart with a statistical process control chart.

The need to understand the process through the analysis of the statistical product control chart tool will be a recurring theme throughout this book.

There are a number of instability detection rules based on laws of probability. The four recommended rules for detecting instability using the Individuals chart may be found in the Appendix "Chapter 2, Instability detection rules for the Individuals control chart."

Capability

The dictionary defines *capability* as "The quality of being capable, capacity, ability." (http://dictionary.com)

We can use simple arithmetic to compare the upper and lower limits of the stable Individuals control chart found in Figure 2.3 to the customer specification and quickly determine if the process is capable of providing the customer product "well within" the customer diameter specification.

For discussion purposes, we will assign a customer diameter specification of 0.5 ± 0.005 to the machined parts of Figure 2.3. The low specification is 0.495, and the high specification is 0.505.

For a critical product characteristic, many companies subscribe to the definition that minimum capability exists when data for that characteristic in question is centered at the nominal specification and the total amount of product variation (the difference of the calculated upper control limit minus the lower control limit of the Individuals control chart) takes up no more than 75% of the customer specification.

When these conditions exist, a simple ratio of total specification to total amount of product variation can be used to determine the capability ratio (CR) for the product characteristic being analyzed:

$$CR = \frac{\text{100\% of Specification}}{\text{75\% of Total Product Variation}} = 1.33$$

A 1.33 capability ratio is the minimum required by most manufacturers and can easily be determined for the process defined by Figure 2.3 using the following arithmetic:

Customer nominal = 0.5
Customer upper specification = 0.505
Customer lower specification = 0.495
Customer specification range is 0.01 total (0.505 – 0.495)
Process is centered at 0.5
Variation = 0.006 total (UCL – LCL; 0.503 – 0.497)

$$\text{Diameter Capability Ratio} = \frac{\text{Total Specification}}{\text{Total Variation}}$$

Diameter Capability Ratio = 0.01/0.006
Diameter Capability Ratio = 1.66

The process described by the control chart illustrated by Figure 2.3 is stable and capable with regard to the product characteristic "diameter." The pattern of data points does not violate the detection rules for instability found in the Appendix, and the CR is 1.66 which exceeds the 1.33 CR that is the minimum required by many companies.

Ford's Contribution

Obviously the term *capability* in the manufacturing world of the twenty-first century comes with its own definition.

The earliest reference I have found to this unique definition of manufacturing capability in American industry can be found in the Q101 Ford Motor Company Quality Manual of the early 1980s.

In this manual, Ford requires all vendors to demonstrate that the critical characteristics identified by Ford take up no more than 75% of the Ford specification when the supplier's process is centered on the nominal value. This would be a CR of 1.33.

$$\text{Capability Ratio} = \frac{\text{100\% of Specification}}{\text{75\% of Product Variation}}$$

$$\text{Capability Ratio} = \frac{100\%}{75\%}$$

$$\text{Capability Ratio} = 1.33$$

For example, consider a Ford requirement for a molded plastic component used in the fuel injection system with a critical width of 1.25 ± 0.01; the full range of acceptable product would be 1.24 to 1.26.

With respect to the width characteristic, in order to qualify as a supplier for the molded plastic component before the revised Q101 requirements were published, suppliers were required to submit samples and an inspection report. The inspection report had to indicate the samples had been measured by the supplier and the width characteristic was within Ford specification on all the samples.

In order to qualify as a supplier for the same part after the imposition of Q101, suppliers were required to submit samples and a statistical product control chart. The control chart had to demonstrate the width measurements were randomly distributed about the centerline indicating process stability. Also when using the Individuals chart, the width of the control chart limits (UCL – LCL) must not consume more than 75% of the total specification when the centerline of the control chart was at the Ford nominal.

For the Ford part previously described, the control chart would have to demonstrate

- Process stability—There should be random distribution of width measurements about the centerline and should not violate any of the four detection rules for instability.
- Process capability—In order to be capable when the product average is at 1.25, the ratio between the difference of the UCL minus the LCL and the full range of the specification must be equal to or greater than 1.33.

Specifically, for this part to satisfy Ford's capability requirement of 1.33, the LCL and UCL of the Individuals control chart with the centerline at 1.25 would have to be (rounding off) 1.246 and 1.253, respectively.

The arithmetic for this conclusion is as follows:

Minimum CR of 1.33 = Total Specification × 0.75
= 0.01 × 0.75 = 0.0075
Control Chart Lower limit = 1.25 centerline – ½ of 0.0075
= 1.25 – 0.00375
= 1.24625
Control Chart Upper limit = 1.25 centerline + ½ of 0.0075
= 0.00375 + 1.25
= 1.25375

For demonstration purposes, the final numbers above are not rounded off, but the control chart limits have been rounded off.

- Control chart centerline at the Ford nominal of 1.25
- Lower control chart limit at 1.246 minimum
- Upper control chart limit at 1.254 maximum

Candidate suppliers competing to produce this part for Ford, under the new rules of Q101, would be required to provide samples and a statistical product control chart with the centerline at nominal, data points randomly distributed about the nominal, and the LCL at 1.246 maximum and the UCL at 1.254 minimum.

Ford's Supplier Base Reaction

The reaction of the supplier base was predictable—outrage at Ford's "arbitrary reduction of specification."

But look at the Q101 requirement from Ford's perspective.

Suppose Supplier XYZ had been producing the product in question for several years with tooling that was machined to accommodate 1.25 nominal width. The amount of width variation resulting from shrinkage, variation in raw material melt indexes, voltage fluctuations, operator technique differences, and so forth, results in the product consuming the entire range of the Ford specification of ±0.01. As long as the process remains stable, the product shipped to Ford will range in width from 1.25 to 1.26, and Ford will receive product within specification.

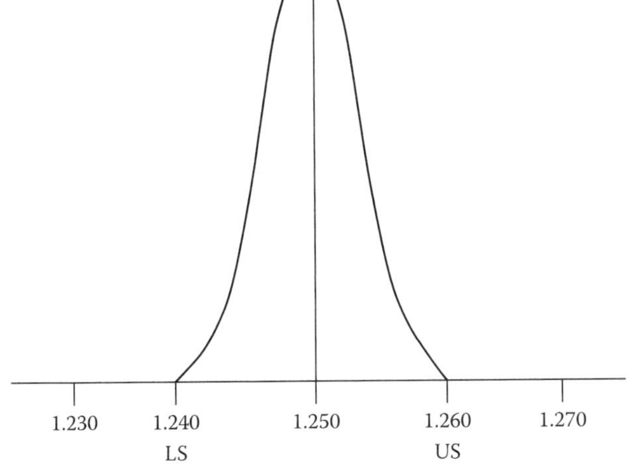

Figure 2.5 A stable process centered at nominal and consuming the entire range of specification. All product falls within Ford width specifications.

Figure 2.5 illustrates this process output in histogram form when the process is stable.

With rare exception, product will be produced within Ford's width specification as long as the process average remains centered at the Ford nominal of 1.25″.

Figure 2.6 illustrates what happens when the entire range of specification is being taken up and the process experiences a period of instability, which, in this case, causes the process average to shift upward. The arrow indicates the amount of shift in the average width due to process instability, and the dotted line curve shows the effect on the process output after the shift.

Ford's Reasoning

The reasoning behind Ford's action was as follows:

- Ford Motor Company had, for years, provided suppliers with specifications. The nominal was the optimum dimension desired for Ford ease

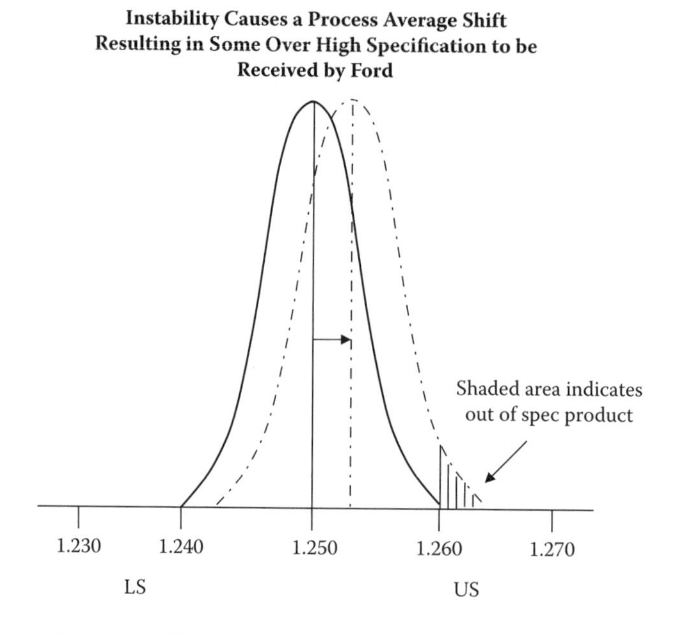

Figure 2.6 Instability causes a process average shift, resulting in some product over high specification to be received by Ford.

of assembly and overall quality. The lower and upper specifications were what Ford was willing to accept, not desirable but acceptable.

- Over the years, the suppliers had become accustomed to consuming the entire range of specification.
- No process is safe from minor process average shifts due to occasional periods of instability.
- Ford would receive under low and over high parts when a supplier's process was consuming the entire range of specification and minor instability was experienced. Out-of-specification parts would result in rejects if found at incoming inspection and slowed production if found on the assembly line.
- If a supplier was consuming less than the entire specification and a minor shift in the average due to instability, there would be a better chance that under low and over high specification product would not be produced and shipped to Ford.
- Therefore, beginning with the publication of Ford Q101, all suppliers were required to demonstrate their capability of providing Ford with product that consumed no more than 75% of the agreed-on specification when the product average was centered at nominal.

Six Sigma

Thus began the need for manufacturers to understand and perform process capability studies and the road to what eventually became known as *Six Sigma.*

Six Sigma is a term coined by Motorola and, in the most basic terms, it requires a supplier to provide product that consumes no more than 50% of the customer specification when the product average is centered at the customer's nominal. Motorola's requirement is considerably more stringent than the original Ford requirement. Six Sigma has evolved to encompass the total quality management philosophies and procedures initially introduced by Dr. Joseph M. Juran in 1964.

A more detailed explanation of the original Six Sigma concept can be found in the Appendix under "Chapter 2, Six Sigma."

CASE STUDY 2.1 ISOPRENE PAPER FEED BELT WIDTH

A manufacturer of copiers designed an isoprene rubber paper feed belt critical to the smooth operation of a specific copier model. The contract to supply this belt was awarded to a New England–based corporation specializing in the manufacture of polymeric office equipment components. For approximately 3 years, the product was provided to the copier manufacturer free of quality problems. The copier manufacturer retired the model and instructed the supplier to place the mold used to manufacture the isoprene belt in storage.

Several years later, the copier manufacturer was in the process of designing a new copier and decided tooling costs could be avoided if the old belt configuration was designed into the new copier. The one difference would be in the belt material; the new belt would have to be manufactured using silicone rubber instead of the original isoprene rubber.

The supplier removed the mold and associated tooling from storage and, as requested, executed a sample production run of 300 molded sleeves that were cut and trimmed to produce 1,200 finished belts. Fifty finished belts were selected at random, in sequence of manufacture, from the beginning to the end of the molding process. Measurements of three critical characteristics were recorded for the 50 samples—belt width, belt thickness, and height of a center bead on the inner diameter of the belt.

Based on the statistical product control chart for the belt width, the inspection department reported the process was stable but not capable.

The 50 belt width dimensions are listed in Table 2.2.

The Individuals product control chart for the 50 sample belts is illustrated in Figure 2.7. There are no violations of the four detection rules for instability; therefore, the process can be deemed as stable.

The customer specification for belt width was 0.75 ± 0.015.

Table 2.2 Belt Width Data

0.748	0.747	0.748	0.758	0.743
0.744	0.742	0.748	0.742	0.746
0.751	0.741	0.757	0.753	0.742
0.756	0.745	0.750	0.755	0.748
0.756	0.746	0.749	0.760	0.750
0.759	0.739	0.747	0.750	0.750
0.739	0.747	0.760	0.747	0.748
0.749	0.748	0.754	0.753	0.761
0.755	0.751	0.762	0.748	0.741
0.745	0.748	0.747	0.754	0.746

Note: Width specification –0.750 ± 0.015.

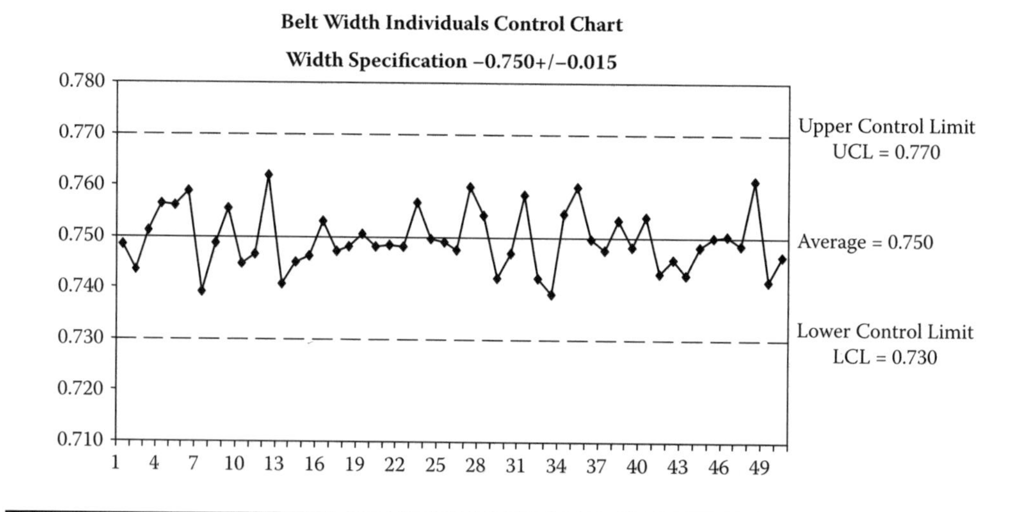

Figure 2.7 Belt width individuals control chart (width specification –0.75 ± 0.015).

Lower specification = 0.735
Upper specification = 0.765

It should be obvious to the reader that, although the process is stable with respect to the belt width, it is not capable. The Individuals product control chart average is at the customer nominal of 0.750, but the calculated control chart limits based on the individual measurements of the 50 samples predicts a 0.730 LCL and a 0.770 UCL. This degree of normal variation when compared to the customer specification indicates a 0.75 capability ratio:

$$CR = \frac{\text{Total Specification}}{\text{Total Product Variation}}$$

$$CR = \frac{0.03}{0.04} (\text{UCL} - \text{LCL})$$

$$CR = 0.75$$

A certain amount of product shipped to the copier company will exceed both the upper and lower customer specifications.

There are two very important points emphasized by this case study:

A process can be stable and yet not capable.

The full range of samples selected randomly from a production run does not necessarily represent the full range of measurements that will be found if all the product were to be measured.

When the results of the study were presented to a group of managers, several representatives from the marketing department insisted the process was capable because all the samples were within the customer specification. Also, it was strongly suggested the belt should go into full production as the customer was behind schedule in its prototype build effort.

Cooler heads pointed out that the data provided by the 50 samples did not represent the total amount of belt width variation. The Individuals control chart limits indicated the total belt width variation was 0.04 (±0.02), which was consuming more than the customer total specification of 0.03 (±0.015). Regardless of what the samples indicated, the copier company would receive some percentage of product oversize and undersize if the process were not corrected.

Based on the data from Table 2.2 and the customer's minimum CR requirement of 1.33, the supplier company would need a specification of 0.027 (rounded off) either side of nominal for its process to be considered capable:

$$CR = \frac{\text{Total Specification}}{\text{Total Product Variation}}$$

$$CR\ (1.33\ \text{minimum}) \times \text{Total Product Variation} = \text{Total Specification}$$

$$1.33 \times 0.04\ (\text{UCL } 0.57 - \text{LCL } 0.53) = \text{Total Specification}$$

$$1.33 \times 0.04 = 0.0532$$

0.0532 = Total Specification required to meet a 1.33 CR with the existing Total Product Variation.

The decision was made to contact the customer and request a blueprint change for the belt width. The requested change to 0.75 ± 0.027 (0.723 to 0.777) would meet the customer's required CR of 1.33. The alternative would be time-consuming and expensive delays that would result from modifications to the existing mold and associated tooling.

A conference call was initiated with the copier company design engineering group.

The immediate reaction of the design engineers was to refuse the requested increase in specification in favor of reworking the tooling. It was pointed out the hard tooling could be reworked at some considerable expense, but the tooling was on consignment to the supplier as it belonged to the copier company. The cost of any rework would be the responsibility of the copier company. The teleconference ended with the customer representatives explaining they needed upper management input. Within the hour, the copier company requested a dozen belts specially processed so as to meet the requested specification range.

Several days later, six belts were specially processed to meet the low requested belt width of 0.723, and six belts were processed to meet the high requested belt width of 0.777.

These belts were shipped to the copier company design engineering group; the belts were installed on test platforms and performed without difficulty. New blueprints were issued with the belt width identified at 0.75 ± 0.03. Several years of subsequent production ensued without customer complaint or return for dimensional problems.

If the process capability study had not been performed, it is reasonable to believe the sample submission would have resulted in the approval of the supplier's production process. And, because the copier company was behind schedule in the release of the new model, it is reasonable to believe the first few shipments of production would have been accepted and installed in copiers.

At some point, the copier company's inspection procedures would have discovered that incoming inspection samples—when compared to the original specification—had oversize or undersize belt widths. This would have resulted in the all too familiar progression of customer complaints, sales returns, increased inspection, internal rejections, rework, scrap, customer quality visits, and so forth.

This one application of the process capability study technique might very well have returned the full cost of educating the engineers and managers in the discipline of SPC and process capability studies. Also, this case study demonstrates that sometimes customer specifications are arbitrary.

Chapter 3

Performing a Variable Process Capability Study

> The prime question—What kind of problem is it? Is it a problem of instability or is it a problem of incapability?
>
> **D.B. Relyea**

Introduction

Many manufacturing processes experience one or both of two categories of defects—variable defects or attribute defects.

A variable defect can be expressed numerically on a continuous scale in terms of inches, millimeters, density, percent elongation, and so forth. An attribute defect is expressed in terms of good or bad, voids or no voids, burrs or no burrs, etc.

This chapter will address the process capability study as it relates to variable data, and Chapter 4 will address the process capability study for attribute data.

In Chapter 2, a definition of the term *capable* was provided in context of the process being centered at the customer nominal. Ford, Motorola, and other companies that require suppliers to demonstrate capability recognize that processes will not always be centered exactly at nominal. Sometimes the process is not centered exactly at nominal due to process conditions, and other times the process is purposely offset from nominal for reasons of economy.

The following case study illustrates how a manufacturer satisfied a customer requirement and became more competitive by deciding to produce product below the customer's nominal and still maintain a 1.33 minimum capability.

CASE STUDY 3.1 INSULATION WALL THICKNESS

A manufacturer of electric power cable used by utility companies all over North America was required by its largest customer to cost reduce a specific product in order to maintain the existing level of business. The customer requested a written plan within 30 days outlining the steps the supplier intended to take to cost reduce the product by 5%.

An analysis of past production records, which included measurement data provided by various operators, revealed the cable insulation—as measured by the thickness of the insulation surrounding the conductive material (wall thickness)—had, for several years, been running at or near the upper specification. Further investigation uncovered the fact that the wall thickness insulation began running on the high end of the customer specification shortly after an inspector rejected a shipment because the insulation wall thickness was below the customer's low specification. It was suggested that perhaps the customer's cost reduction requirement could be realized by reducing the average wall thickness of the insulation material and placing it at the customer's nominal specification.

The point was made that it would be imprudent to lower the average insulation wall thickness to nominal without first determining the amount of normal variation of insulation wall thickness that the process was capable of maintaining.

This particular company had no history of performing process capability studies, even though several of their engineers and technicians had attended statistical process control seminars. The rationale for not performing process capability studies was their customer base had never made any requests for control chart results and customers had never mentioned any process capability requirements. Historically, the primary concern of the customer base had been that they not receive any product that was under the low specification for insulation wall thickness.

It was suggested that a process capability study be performed immediately in order to determine the current average and normal variation of the insulation wall thickness of the product currently being produced which was the product the customer had requested be cost reduced.

A measurement process was quickly qualified, and the process capability study proceeded. Measurement process qualification will be discussed in detail in Chapter 6.

The process capability study was conducted over a period of 24 hours without interfering with normal production. A sample was selected from the beginning and the end of each reel of cable produced during the study. Figure 3.1 illustrates the Individuals product control chart representing insulation wall thickness variation for the product produced during the study.

The Individuals chart indicates the wall thickness characteristic to be stable (no violations of the four instability detection rules) with the average wall thickness of 0.283″, and the control limits indicate a total amount of wall thickness variation equal to 0.006, 0.003 either side of the wall thickness average.

Figure 3.2 shows the same data depicted in the form of the normal curve compared to the customer specification of 0.270 ± 0.015 (0.255 to 0.285).

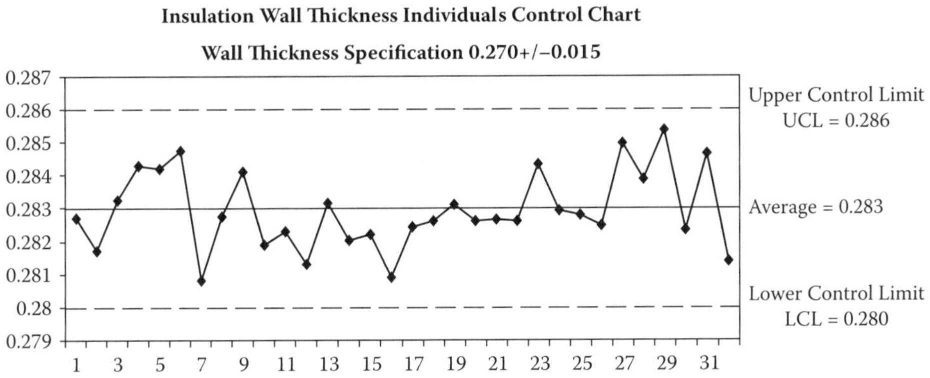

Insulation Wall Thickness Individuals Control Chart

Wall Thickness Specification 0.270+/−0.015

Upper Control Limit UCL = 0.286

Average = 0.283

Lower Control Limit LCL = 0.280

Figure 3.1 Insulation wall thickness individuals control chart (average = 0.283, lower control limit [LCL] = 0.280, upper control limit [UCL] = 0.286, wall thickness specification 0.270 ± 0.015).

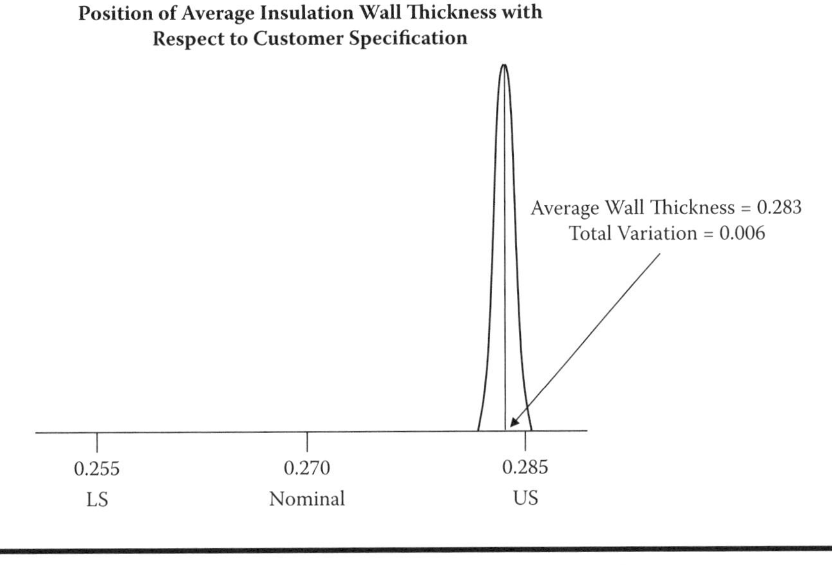

Figure 3.2 Position of average insulation wall thickness with respect to customer specification.

The reader should not be dismayed at the narrowness of the normal variation with respect to the amount of customer specification. The vast majority of process capability studies I have been involved with over the past 35 years, when performed with a qualified measurement process, have indicated similar results as are depicted in Figure 3.2. I am convinced the majority of the world's manufacturers do not appreciate how very capable their processes are because of poor measurement processes, unnecessary operator process adjustments, and use of the wrong method to determine standard deviation often artificially inflate the variation of measured data. These topics will be covered later in the book.

There is also the very real fact that most specifications are rather arbitrary and are created with little or no knowledge of the supplier's process capability.

Figure 3.2 clearly indicates potential for a significant reduction in raw material cost if the process were adjusted to meet the customer's nominal specification.

In Chapter 2, the term *capability ratio* (CR) was used to describe the capability of processes when the average of the product characteristic was positioned at the nominal specification. When the product average is at the nominal specification, both the lower and the upper specifications are an equal distance from the low end and high end of the product variation, respectively.

The arithmetic for the CR is

$$CR = \frac{\text{Total Specification}}{\text{Total Variation}}$$

Figure 3.3 represents the insulation wall thickness if it were positioned at the customer nominal providing a CR of 5.

Note that, in terms of sigma, the lower specification (LS) is the same distance from the low end of the product variation as the upper specification (US) is from the high end of the product variation. This balance of distances allows the use of CR.

The CR for Figure 3.3 is determined as follows:

$$CR = \frac{\text{Total Specification}}{\text{Total Variation}}$$

$$CR = \frac{0.03}{0.006}$$

$$CR = 5$$

Initially, management did, in fact, direct the standard operating procedure (SOP) be modified to reduce the raw material usage by shifting the product wall thickness average to coincide with the customer's nominal. This would significantly reduce the cost of raw material and come close to satisfying the customer cost reduction requirement without violating the concept of maintaining the internally imposed 1.33 CR minimum. Please remember the

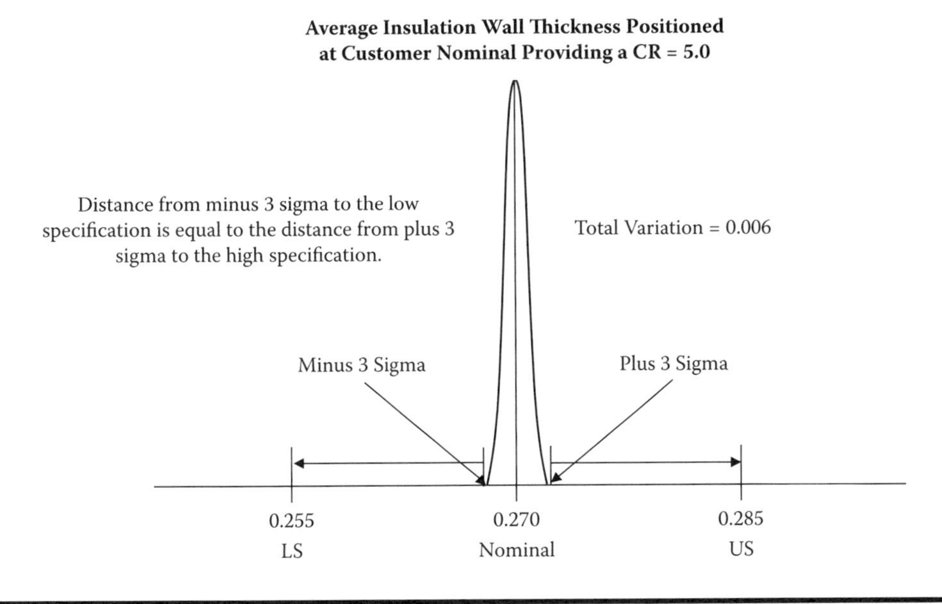

Figure 3.3 Average insulation wall thickness positioned at customer nominal providing a capability ratio of 5.

customer had no capability requirements, but the cable company was being guided to abide by the industry-wide minimum CR of 1.33.

It was further suggested that a refined concept of CR should be considered which would allow for even more savings. Management was introduced to the concept of capability index (Cpk).

Capability Index (Cpk) versus Capability Ratio (CR)

The Cpk is the same concept as the CR. The only difference is that we use Cpk when the product average is not the same as the customer nominal specification.

The arithmetic for CR develops one number such as 1.33 to describe the capability of the process. The arithmetic for Cpk develops two numbers:

- The capability index for lower specification (CPL) describes the capability of the process in relation to the lower specification.
- The capability index for upper specification (CPU) describes the capability of the process in relation to the upper specification.
- One number, the lower number of CPL and CPU, is used to describe the capability of the process, and it is termed *Cpk*.

Many people today use the term *Cpk* even when the product characteristic is centered at nominal. Also, it is standard practice to communicate to the customer one Cpk value when there is a difference between CPU and CPL; the Cpk value communicated would be the lower of the two available numbers. Going forward, these two practices will be used in this book.

The question for management now becomes how to determine the Cpk when the average is not on nominal while maintaining the industry standard minimum Cpk of 1.33.

The answer is rather simple. Because the center of the normal curve will no longer be on the nominal, the distance between product average and the lower and upper specifications will no longer be equal. If the distances are not equal, we now have two numbers to describe the capability of the process. These numbers will take into account

- The customer's lower specification, nominal, and upper specification
- The product average
- The amount of wall thickness variation about the product average

One number will describe the distance from the low end of the product variation to the lower specification, and the other number will describe the distance from the high end of the product variation to the upper specification. These numbers are called the CPL and the CPU.

Convention mandates that if either one of these two numbers is less than 1.33, we cannot call the process capable.

If management decided to change the SOP and place the wall thickness average at 0.26, the arithmetic to determine the CPL and the CPU would be as follows:

CPL = Product Average – Lower Specification ÷ ½ Total Variation
CPL = (0.26 – 0.255) ÷ ½ × (0.006)
CPL = 0.005 ÷ 0.003
CPL = 1.66
CPU = Upper Specification – Product Average ÷ ½ Total Variation
CPU = (0.285 – 0.260) ÷ ½ × (0.006)
CPU = 0.025 ÷ 0.003
CPU = 8.33

Both the CPU and CPL are greater than 1.33, so we can state the process is capable. It is capable of experiencing minor process average changes while continuing to produce product within specification.

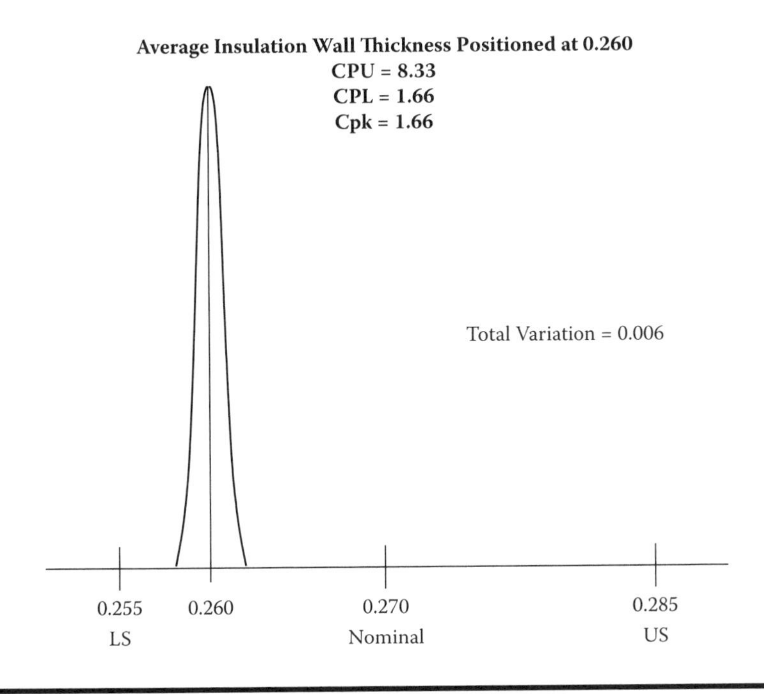

Figure 3.4 Average insulation wall thickness positioned at 0.26 (capability index for upper specification [CPU] = 8.33, capability index for lower specification [CPL] = 1.66, capability index [Cpk] = 1.66).

The lower of the two numbers is referred to as the Cpk. The Cpk for this process with the product average set at 0.26 is 1.66 (Figure 3.4). This process is capable, and a significant amount of raw material will be saved over time.

Management decided to revise the SOP to require the process to produce insulation wall thickness at a 0.26 average. The savings in raw material more than satisfied the 5% cost reduction required by the customer. The balance of the savings went directly to the cable company's bottom line.

A series of process capability studies was planned for similar product.

The case study just offered along with the molded belt case study provided in Chapter 2 should indicate to the reader that the process capability study is a valuable tool that can be applied in various ways to improve the competitiveness and profitability of an organization.

The molded belt study in Chapter 2 is an example of using the Cpk concept to ensure the smooth transition into production of a new product. The insulation wall thickness example is a demonstration of optimizing a well-established process by minimizing raw material usage while maintaining a safe margin within customer requirements. The same requirement to demonstrate stability and numerically describe capability in terms of Cpk should be imposed on a manufacturer's supplier base. It is a safe assumption that stable and capable raw material will only serve to make a manufacturing facility more competitive. Also, the capability study can and should be used on a daily basis by shop floor associates to solve production problems.

The process capability study technology literally applies from one end of a manufacturing facility to the other.

Using Process Capability Studies to Solve Shop Floor Problems

The terms *process stability* and *process capability* form the basis for every problem-solving exercise that takes place within an organization.

Problem solving begins with a clear definition of the problem, but many times, problems are ill defined. Most often, problems are defined in terms of the symptoms such as "we have a diameter problem on line #17," or "Acme called and our units are too long."

Generally speaking, product problems on the shop floor or at the customer or supplier fall into one of two categories. The problem can be one of *instability* or of *incapability*. Specifically, the problem exists because the product average has shifted or because the product has too much variation about a consistent average.

Figure 3.5 illustrates the two categories of product problems that can exist. Figure 3.5a indicates a process that is experiencing instability. Note the process average shift represented by the dotted line normal curve. Also note the instability, in this case, is not producing product outside customer specification—yet. Regardless of the degree, it is always prudent to identify the root cause of the instability. Figure 3.5b indicates a process that is stable but incapable of producing all product within the customer specification.

As stated earlier, the first step in problem solving is to clearly define the problem as instability or incapability. And instability or incapability can only be quantified by means of a process capability study.

Figure 3.6 is a flow diagram outlining the recommended method to facilitate a process capability study.

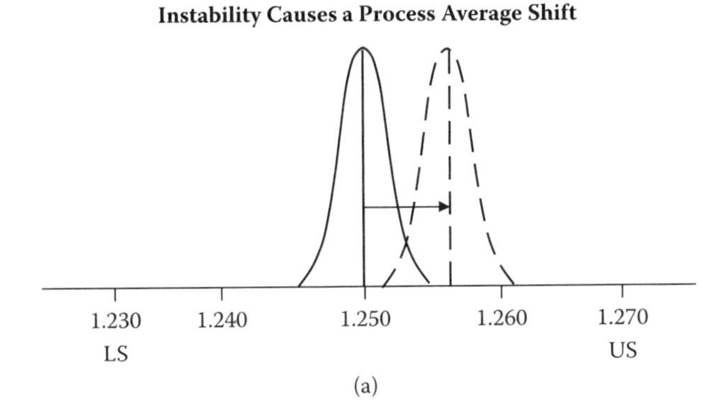

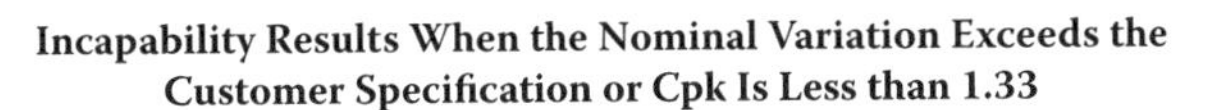

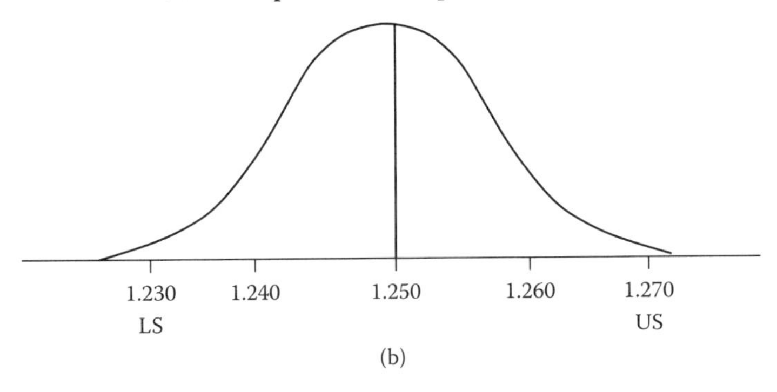

Figure 3.5 (a) Instability causes a process average shift. (b) Incapability results when the normal variation exceeds the customer specification or the capability index (Cpk) is less than 1.33.

Shop Floor Interface

A successfully executed process capability study depends largely on operator buy in. To this end, every precaution must be taken to ensure the operator willingly participates in the study fully convinced that it is the process and not the operator which is being evaluated. Before the study begins, it is a good idea to invite the supervisor of the area and the operators who will be engaged in the study to the training room for a brief review of normal variation as it relates to critical characteristics. At this time, the operators should be requested to refrain from making adjustments during the study; this is often a difficult request for many operators to accept. Witness how often machine operators will select and measure a random sample from the output of a process and, if the measured characteristic is not exactly at the

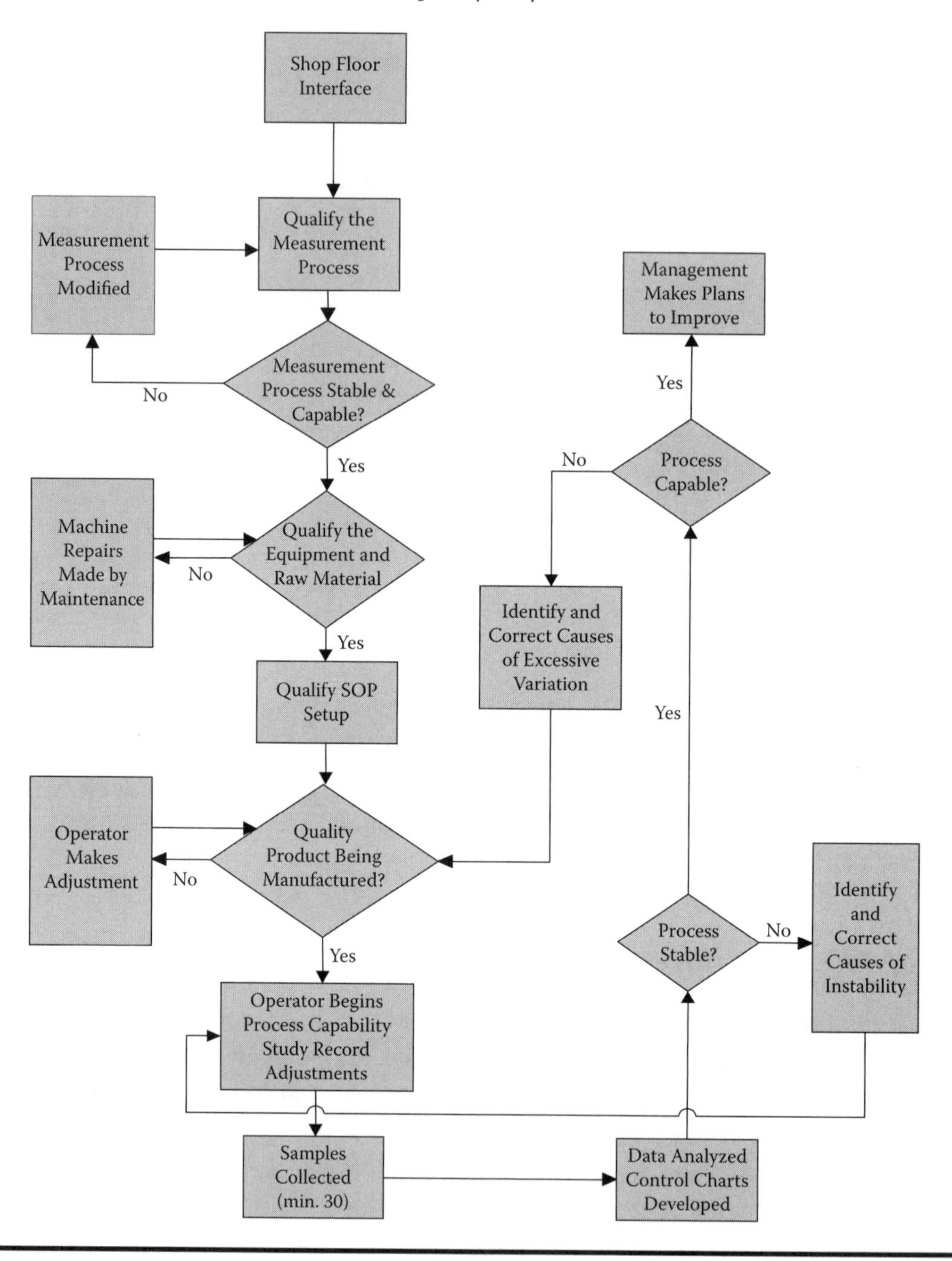

Figure 3.6 Process capability study flowchart.

desired value, the operator will make an adjustment to the process. This is an indication of the widespread mistaken belief that it is within the ability of the operator to minimize or even eliminate normal variation by making adjustments. This belief is so ingrained that it is a good idea to ask the operator to make notes if he or she feels absolutely compelled, at some time during the study, to make an adjustment. The notes should include the process parameter that was adjusted as well as the approximate time the adjustment was made. This information will be useful during the analysis of the process capability study results. Many times, process average shifts identified on the control chart during the analysis are matched with adjustments the operators believed were necessary.

Every effort should be made to minimize outside influence on the process during the process capability study, especially the presence of numerous individuals not normally on the shop floor hovering over the operators. It is best to assign one person to interface with the operators several times during the study to address any concerns they may have regarding the procedure.

Qualify the Measurement Process

The purpose of a process capability study is to determine if the manufacturing process is stable and capable by measuring the critical product characteristic of interest. The product measurements are then used to develop a statistical product control chart. By analyzing the data patterns on the control chart, we determine if the process is stable. If the process is stable, we then use the data to determine a Cpk value in order to determine the degree of capability.

It is *impossible* to determine if a manufacturing process is stable if the measurement process used to measure the product is not stable. This is because an unstable measurement process will make a perfectly stable manufacturing process appear to be unstable. Unfortunately, unstable measurement processes in manufacturing are common.

If the measurement process is determined to be stable, we must then ensure the measurement process is capable. Incapability exists with respect to the measurement process when the amount of measurement variation is inappropriately high when compared to the amount of product variation.

All too often, a stable but incapable measurement process will falsely indicate the manufacturing process is generating excessive variation in a critical product characteristic because the measurement variation is being added to the product variation. Many manufacturing processes believed to be incapable (Cpk < 1.33) are found to be more than capable (Cpk > 1.33) once the measurement process incapability is corrected.

Performing a measurement process analysis (MPA) will be fully discussed in Chapter 6.

Qualify the Equipment and Raw Material

In a perfect world, shop floor personnel would establish process parameters according to the SOP and introduce raw material to the equipment confident in the knowledge the critical raw material characteristics would be stable from batch to batch and capable (Cpk > 1.33). Anyone familiar with the typical manufacturing process knows the shop floor is not a perfect world, and sometimes the equipment is not in perfect working condition, and the raw material may not be as prescribed by the SOP.

Commercial airline pilots often perform a "walk around" on their aircraft before takeoff. The purpose of the walk around is to perform a visual inspection of the exterior just to ensure the structural integrity of the aircraft before it becomes airborne.

It is wise to perform a walk around on equipment with the operator and supervisor before a process capability study is performed.

After more than 40 years of being associated with manufacturing organizations, I can safely say I have been involved in hundreds of process capability studies. I estimate about 30% of the process capability studies I have been involved with required equipment repairs before the study could proceed. Some observations included

- A pressure gauge was inoperable. The face glass was missing and the indicator was below zero in spite of the fact that the screw (auger) was rotating with raw material in the barrel. The supervisor had issued a work order several months prior to the walk around, but the maintenance associate who came to replace the gauge had been called away for a "more critical repair." The operator was instructed to proceed with production until the gauge was replaced. Apparently the gauge repair was lost in a bureaucracy. When asked, the operator stated she could determine the head pressure by the "sound of the screw."
- A tensioning device was inoperable; the operators compensated for the condition by leaning a pallet against the "give up."

- Operators had jumpered out a safety interlock in order to facilitate increased production.
- A thermocouple was held in place by masking tape.
- A dietary scale was brought from home to take the place of a broken gram scale.

A walk around before a process capability study is recommended.

A review of raw material condition is also suggested. A few raw material anomalies identified over the years include

- Raw material used in a compression molding operation was preheated which was not called for in the SOP.
- First-in, first-out protocol was violated due to general agreement among the shop floor personnel favoring one supplier's raw material over another.
- Contraband lubricants were used based on the belief the prescribed lubricant was not effective.
- Contrary to the SOP, batches of raw material were being mixed by the machine operator instead of in the confines of the chemical room.
- Raw material was drawn from a warehouse in Midwestern Canada, in February, and was not allowed to attain room temperature before use.
- The supplies of raw material were used up shortly after the study began.

A review of the raw material situation and availability before launching a process capability study is recommended.

The following case study is an example of the variable process capability study as it applies to solving a shop floor problem.

CASE STUDY 3.2 CRIMPED TERMINAL DIAMETER CUSTOMER RETURNS

A U.S. automotive company contracted with an electronic component manufacturer to provide a 10-inch-long, laminated flexible assembly designed to connect the automobile radio to the power, speakers, and ancillary equipment terminated in the dashboard. Four individual lead wires extending beyond the laminated portion of the assembly served to provide the necessary electrical connection to the bus bar in the dashboard.

The electrical connections were formed by mechanically crimping terminals to the four lead wires. The diameter of the finished crimped terminals was critical in order to provide a proper fit when mated to the dashboard bus bar. Almost from the very first shipment, the supplier received sales returns and complaints due to oversize and undersize terminal diameters.

The oversize terminals would not easily fit into the precisely designed bus bar, which slowed production, and the undersized terminals fell out of the bus bar after points of sales, resulting in warranty claims.

The specification for each of the four finished terminals was 0.042 ± 0.003; 0.039 lower specification and 0.045 upper specification. At the manufacturer, the assembly process consisted of each of the four lead wires being individually inserted into a die that was automatically "loaded" with a terminal from a feed roll, the operator would then activate a pneumatically powered die, and the lead wire would be crimped with a terminal. Then the next lead wire would be inserted, and the process would be repeated. Three dies were active on each of three shifts for a total of nine process streams.

Management's response to the initial customer complaints and returns was to have the die operators use handheld micrometers to measure the four crimp diameters on one finished part for every 25 assemblies. If an out-of-specification crimp was discovered, the die or the pneumatic pressure or both would be adjusted.

The returns and complaints continued.

The next corrective action was to have a sample of 125 assemblies selected from each lot of approximately 2,300 finished parts. The diameters of four terminals on each of the 125 assemblies were measured with a micrometer, and if any crimped diameters were out of specification, the entire lot was sorted 100%. Many parts were scrapped off, and the remainder of the parts was shipped.

The returns and complaints continued.

The machine operator sample inspection was discontinued, and part-time high school students were hired to inspect 100% every finished crimped terminal. Rejects piled up at considerable cost.

The returns and complaints continued.

After several months of returns and complaints, I was invited to participate in helping to solve this problem the day after the automotive customer informed the supplier that continued rejects would result in the business being moved to a competitor.

The morning of day one of the new problem-solving effort began with an initial meeting with management and several supervisors and engineers. Discussions revealed the crimping process had been released to production from product design without any measure of stability or capability. When the process was initially set up on the shop floor, a handful of samples had been selected and measured to someone's satisfaction, and the process was turned over to manufacturing. Process capability studies were not part of this company's culture in spite of the fact that most of the engineering staff had attended statistical process control seminars.

The process capability study concept was briefly reviewed, and management quickly accepted the suggestion that a study be performed to determine if the crimp diameter problem was instability or incapability.

Before we could begin, it was necessary to qualify the measurement process with a proper MPA. The group contended it would not be necessary to perform a measurement study, because there was an extensive calibration program that ensured all micrometers were accurate.

It was pointed out that accuracy of a measurement device is necessary, but precision discrimination and reproducibility of a measurement process are also important.

Accuracy—Accuracy is comparison to a known standard. Accuracy of a micrometer is determined by measuring a gauge block. If the gauge block is certified to be 0.25 and the micrometer registers 0.25, the micrometer is deemed to be accurate. In this case, however, the calibrated micrometers were being used to measure thin-walled, fairly compressible terminals. When an operator measured a crimped terminal, the pressure she applied with the calibrated micrometer might compress the terminal and generate a false reading due to a lack of precision.

Precision or repeatability—When discussing measurement processes, we must consider precision that is the ability of the measurement process to measure the same part several times and record results with a minimal amount of variation. Excessive imprecision can result in an unstable measurement process.

Discrimination—Discrimination is the ability of a measurement process to tell the difference between samples made during periods of normal variation. In broad terms, if the measurement variation is equal to or greater than the amount of the normal variation of the product being measured, the measurement process will not be able to adequately tell the difference between some samples, and this could artificially narrow the control chart limits and might lead to false indications of process instability.

Reproducibility—This is the ability of a measurement process to minimize variation of measurements due to operator technique differences.

On the afternoon of day one, the micrometer measurement process was analyzed and disqualified due to excessive imprecision causing measurement process instability. The micrometer measurement process used by production, engineering, and quality assurance could not be used for purposes of performing a process capability study. Then late in the day, a noncontact measurement process was selected and qualified. The new qualified measurement process was a machinist microscope, which would adequately measure the diameters of the crimped terminals without coming into contact and perhaps distorting the terminals.

Once again, the concepts of the measurement process mentioned here, including a step-by-step procedure on how to qualify a measurement process, will be fully discussed in Chapter 6.

On the morning of day two, the three die operators who operated Machine 1 on the three shifts were assembled in the training room and received a 45-(minute) overview of normal variation as it pertains to crimped terminal diameters. The third-shift operator had been requested to stay over, and the second-shift operator had been asked to come in for the short meeting, They were also informed that the first-shift die stamp operator would be requested to set up the process and run the entire shift without making any adjustments. The second- and third-shift operators were also requested to produce product during their respective shifts without making adjustments. The operators were also asked to segregate one sample roughly every hour and mark the samples from 1 to 24.

The next morning we had 25 samples—one more than requested—that were measured by a technician using the machinist microscope. The results can be found in Table 3.1.

Because each assembly consisted of a natural subgroup of four lead wires, the statistical product control chart technique used in this instance was the X Bar and R chart. The X Bar chart records the averages of the four diameters measured for each of the assemblies that were sampled during the study. The R chart records the ranges between the four measurements of each of the assemblies sampled during the study.

For instance, the average of the four measurements in the first subgroup is

	1
	0.0435
	0.0422
	0.0428
	0.0432
Sum	0.1717

$$\text{Average} = 0.1717 \div 4 = 0.0429$$

The range of the first subgroup is equal to the largest diameter minus the smallest diameter in the subgroup. The range of subgroup 1 is

$$\text{Range} = 0.0435 - 0.0422 = 0.0013$$

The calculations for the centerlines and control limits of the two charts can be found in the Appendix under "Chapter 3, X Bar and R chart calculations."

The two charts resulting from the study are illustrated in Figure 3.7.

A second meeting was held with the operators, engineers, and the supervisor of the area for the purpose of reviewing the results of the study.

The initial reaction when the data were projected on a screen was shock that all the data fell well within the customer specification. The largest crimp diameter in the sample was 0.0438, and the smallest crimp diameter in the sample was 0.0412; all were well within the customer specification of 0.039 to 0.045.

After a brief discussion, general agreement was reached that the improved performance was probably due to measuring the crimped terminal diameters using the noncontact method employed with the machinist microscope. Subsequent confirmation proved the unqualified micrometer measurement process was primarily responsible for the customer complaints and returns.

Table 3.1 Crimp Diameter Data

1	*2*	*3*	*4*	*5*
0.0435	0.0428	0.0438	0.0431	0.0422
0.0422	0.0420	0.0424	0.0423	0.0412
0.0428	0.0415	0.0428	0.0414	0.0426
0.0432	0.0422	0.0422	0.0422	0.0421
6	*7*	*8*	*9*	*10*
0.0420	0.0422	0.0425	0.0422	0.0415
0.0425	0.0431	0.0415	0.0419	0.0420
0.0435	0.0433	0.0431	0.0428	0.0422
0.0424	0.0427	0.0423	0.0423	0.0421
11	*12*	*13*	*14*	*15*
0.0419	0.0424	0.0424	0.0420	0.0420
0.0423	0.0420	0.0419	0.0417	0.0414
0.0426	0.0425	0.0422	0.0434	0.0422
0.0421	0.0420	0.0419	0.0428	0.0415
16	*17*	*18*	*19*	*20*
0.0426	0.0422	0.0420	0.0421	0.0421
0.0423	0.0425	0.0420	0.0423	0.0424
0.0428	0.0430	0.0424	0.0421	0.0420
0.0422	0.0421	0.0420	0.0419	0.0422
21	*22*	*23*	*24*	*25*
0.0419	0.0419	0.0425	0.0429	0.0418
0.0414	0.0422	0.0421	0.0414	0.0425
0.0420	0.0426	0.0421	0.0419	0.0421
0.0422	0.0421	0.0418	0.0421	0.0422

I cannot overemphasize the importance of qualifying the measurement process before performing analysis on the manufacturing process.

The X Bar and Range charts were evaluated next.

The rules for detecting instability on the X Bar can be found in the Appendix under "Chapter 3, Instability Detection Rules for the X Bar Chart."

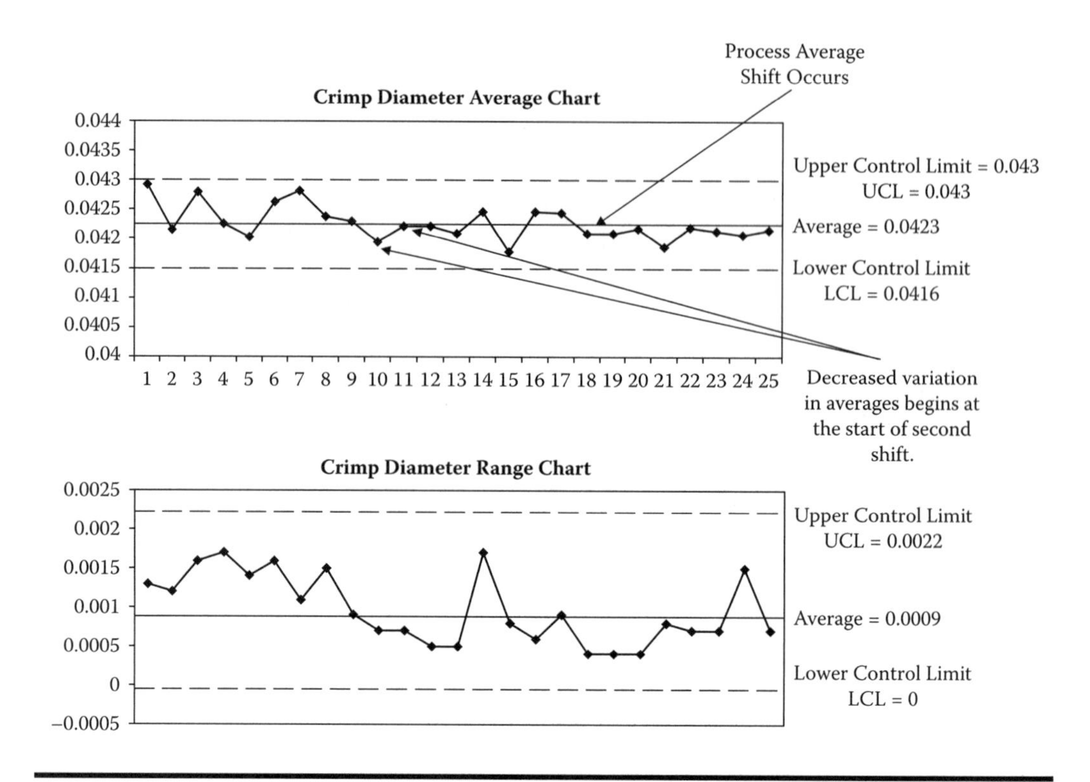

Figure 3.7 (a) Crimp diameter average chart. (b) Crimp diameter range chart.

Please note on the X Bar chart of Figure 3.7 that rule 4 is violated with the last eight averages falling below the centerline; a high probability that instability occurred was coincident with the start of the third shift.

Instability rules notwithstanding, it appears, subjectively, that in addition to a process average shift occurring at the beginning of the third shift, there was also a reduction in overall variation starting at about the beginning of the second shift. Figure 3.7 notes the amount of variation from average to average on first shift compared to the lesser amount of variation from average to average on second and third shifts.

The group agreed that something in the process was changing, and it would be in the best interest of the company to identify the cause of instability and the cause of the apparent reduced variation.

Please note that in spite of the fact that all the samples were well within customer specification, the group believed it was necessary to identify and remove the unknown cause of instability.

Because the operators assured the group they had made no adjustments to the process during the study, we explored other possibilities for the identified changes. Raw material change was discounted, as were electrical and pneumatic fluctuations, ambient temperature changes, equipment malfunctions, and so forth. After some considerable probing, we discovered that the second- and third-shift operators, at the beginning of their shift, manually backed up the feed roll to remove slack going into the die and then tightened up the knurled knob that kept the feed "tight with a minimum of slack." They continued to maintain their "tight" feed throughout the shift by adjusting the knurled knob.

When reminded they were not supposed to make any adjustments to the process during the study, they insisted that what they did was not an adjustment, "it was part of their setup."

Obviously, the second- and third-shift operators were making adjustments and, by doing so, were increasing the tension on the feed roll and subsequently changing the process average and the amount of overall variation. It was time to celebrate, for the group just learned a great deal about the process parameter that contributed to product variation.

Please remember this process consisted of three die machines running on three shifts providing nine process streams. Interviews with the other six operators revealed each one had their own preferred "slack setting," and they adjusted their die machines accordingly. Nine operators constantly adjusted the processes, causing shifts in the process average and changes in the amount of variation, all while measuring the results with an inadequate measurement process. It is not surprising the customer had quality problems.

Figure 3.8 represents the statistical product control X Bar and Range charts developed over two shifts after a constant tensioning device had been installed.

When using the X Bar and R chart for analysis, one sigma of the total amount of variation of the product being studied is determined by dividing the average range (R Bar) by the constant d_2. Table A.1 (Appendix) indicates the constant d_2 for a subgroup size of four (each range was determined by subtracting the smallest measurement from the largest for each subgroup of four) to be 2.059.

One sigma of crimped terminal diameter total variation is

$$\text{One sigma} = \text{R Bar (from Figure 3.8)} \div 2.059$$
$$\text{One sigma} = 0.001 \div 2.059$$
$$\text{One sigma} = 0.0005$$

Because the process average is coincident with the customer nominal, we can use the CR calculation:

$$CR = \frac{\text{Total Specification}}{\text{Total Product Variation}}$$

$$CR = \frac{\text{(Upper Specification} - \text{Lower Specification)}}{\text{Total Product Variation}}$$

$$CR = (0.045 - 0.039) \div (0.0005 \times 6)$$

$$CR = 0.006 \div .003 = 2$$

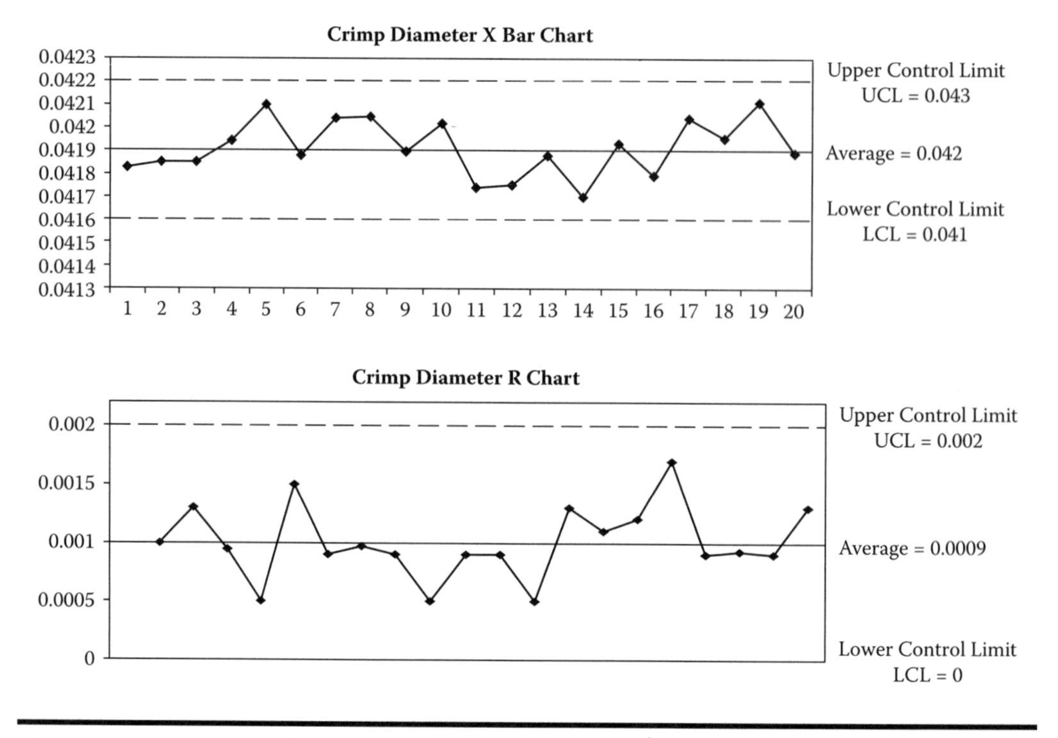

Figure 3.8 (a) Crimp diameter X Bar chart. (b) Crimp diameter R chart.

The report to the automotive customer would indicate the process is stable and capable with a Cpk of 2.

The next step in the evolution was to evolve the data collection of crimped terminal diameters for the statistical product control charts of X Bar and Range to a statistical process control chart for tension.

This concept will be discussed in Chapter 8.

One measurement process analysis and one process capability study performed and analyzed in less than 48 hours led to problem identification and solution—a very inexpensive exercise when compared to the months of adjustments, inspection, sorting, scrap, and complaints.

The most distressing aspect of this obvious success story is the fact that the very next time the electronic component manufacturer encountered a shop floor problem, the process capability study protocol was not employed. I cannot explain this phenomenon of not using a new, yet proven technique to solve problems, but I encounter it over and over.

However, there is one foolproof way for the process capability study protocol to become a part of a company's culture, and that is for members of management to ask one simple question when presented with a "problem": *What kind of problem is it—instability or incapability?*

The appropriate answer cannot be offered unless a process capability study had been performed.

Chapter 4

Performing an Attribute Process Capability Study

There are no guarantees in life, only probabilities.

Unknown

Introduction

The process capability study flowchart found in Figure 3.6 applies to an attribute as well as a variable process capability study.

As mentioned in Chapter 3, attribute defects are those defects of a go or no go nature. When a molded product exhibits "voids" as opposed to the desired "no voids," it is suffering from an attribute defect. Attribute defects associated with manufactured product can be more problematic and expensive than variable defects.

In most cases, discussions at new client companies begin with concerns regarding variable defects such as quality and scrap problems due to out-of-specification crimp diameters. But eventually, someone gives voice to concerns regarding some attribute defect. One common thread that seems to run through every initial discussion regarding attribute defects is the frustration with the perception that attribute defect "seems to randomly come and go." Discussions concerning attribute defects are often interspersed with comments such as "samples from one lot will have no defects and the very next lot will have a defect level of 15% and nothing changed between the two lots … except the percentage of rejects."

There is a general perception in manufacturing that attribute problems are intermittent, and problems that can be measured on a continuous scale—with a micrometer or tensile tester—are with us all the time, just sometimes at an unacceptable level of high or low measurement. This false impression is likely due to the nature of the two methods of inspection—go or no go versus some sliding scale of measured results.

Bead Box Effect

The quickest way to put attribute defect problems into perspective is with a simple bead box demonstration. For over 30 years, I have carried with me a small wooden box, a plastic bag containing 2,000 multicolored beads, and a plastic paddle with 100 recesses molded into the surface large enough to accommodate a population of 100 randomly chosen multicolored beads.

Shortly after the beginning of a discussion regarding a particular attribute defect, I break out my bead box and deposit into it the 2,000 multicolored beads. I ask the group I am working with to consider the population of beads in the box to be a first-shift operator's output on Monday. I point out that 5% (100) of the 2,000 multicolored beads are yellow, and for purposes of demonstration, the yellow beads are to be considered as visually defective product. After stirring up the beads, I use the paddle to select a random sample of 100 beads and show my audience the number of yellow beads in the sample. The beads from the first sample of 100 are deposited back in the box, the contents stirred once again, and a second sample, representing the output of Monday's second-shift operator, is selected with the paddle. Again the results are shown to the audience, and the process is repeated for Monday's fictitious third-shift operator.

Table 4.1 illustrates the results of 30 random samples selected from the bead box at one client location, each sample consisting of 100 beads with the corresponding percent defective recorded for each sample. For discussion purposes, we will consider the Table 4.1 data to represent the output of three shifts spanning 10 consecutive workdays. Note the percent defective column ranges from 2% (two yellow beads in the sample of 100) to 14% (14 yellow beads in the sample of 100).

Table 4.1 Bead Box Results

Sample Number		*Shift*	*Percent (%) Defective*	*Sample Number*	*Shift*	*Percent (%) Defective*
Monday	1	1	2	16	1	8
	2	2	2	17	2	8
	3	3	2	18	3	4
Tuesday	4	1	6	19	1	2
	5	2	8	20	2	2
	6	3	14	21	3	4
7		1	6	22	1	4
8		2	8	23	2	8
9		3	2	24	3	6
10		1	4	25	1	10
11		2	6	26	2	6
12		3	4	27	3	2
13		1	6	28	1	6
14		2	10	29	2	2
15		3	6	30	3	8

Percent Defective (P) Chart

In previous chapters, it was demonstrated how Individuals charts and X Bar and R charts are useful in analyzing the normal variation of variable data such as the width of paper drive feed belts and the diameter of crimped terminals. The normal amount of variation for attribute data can be analyzed by means of a percent defective chart.

The percent defective data follow a binomial distribution that can be visually similar to the normal distribution but requires different arithmetic that will be found in the Appendix under "Chapter 4, Percent Defective (P) Chart Calculation."

Figure 4.1 represents the percent defective chart (P chart) developed using the data from Table 4.1. The centerline on the chart is the average percent defective for the 30 100-piece samples. Note the average of the samples

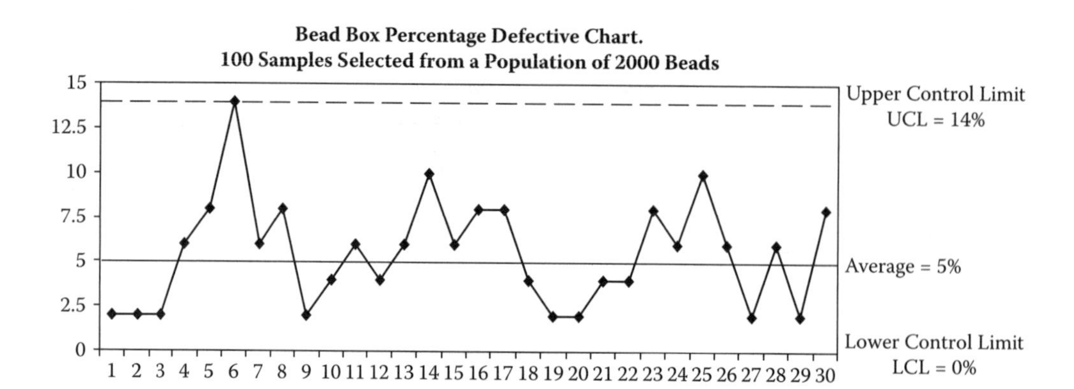

Figure 4.1 Bead box percent defective chart. One hundred samples selected from a population of 2,000 beads.

(5%) is identical to the actual percent defective (percent of yellow beads) in the box. The upper control limit (UCL) of the chart is 14%, and there was, in fact, one random sample of 100 beads that had 14 yellow beads. The lower control limit is 0% defective, and in many demonstrations, zero yellow beads are found in the results of 30 random samples.

The pattern on the P chart indicates the bead box process is stable. This makes sense because the sample beads were returned to the box each time a sample was drawn, and assuming no one added yellow beads to or subtracted yellow beads from the box during the demonstration, the bead box process would remain stable with a constant 5% defective.

The instability detection rules when using the P chart can be found in the Appendix under "Chapter 4, Instability Detection Rules for the P Chart."

When I begin the bead box demonstration, I take great pains to remind my audience the number of yellow beads in the box represents 5% of the total number of beads, and yellow beads are to be considered defective product. The first sample in Table 4.1 is 2% defective (2 yellow beads in the 100 bead sample). After I drew this sample, I held up the paddle and asked my audience the following question: "Based on this sample, what percent defective did the first-shift operator produce?" The response from several members in the group was "2%." I waited, and no one in the group of 17 individuals consisting of a mix of middle managers, engineers, supervisors, and operators contradicted the "2%" comment.

I reminded everyone the number of yellow beads in the box was 5% of the total, and, yes, the sample indicates the first-shift operator produced 2% defective, but we *know* the box contains 5% defective product. At this juncture, the group understood the trap they had fallen into. This particular group of 17 individuals, all employees of a major North American manufacturing facility, is the rule rather than the exception regarding the response to my question.

Acceptable Quality Level

Military Standard 105 Sample Plans for Attribute Defects (Mil 105) was an integral part of quality control decision making for a better part of 50 years. The standard required a customer and a supplier to agree on an acceptable quality level (AQL) for supplied product. For instance, if a customer and supplier agreed on a 1% AQL, they would choose a sampling plan that would provide a certain confidence that, over time, the customer would receive, on average, product that was no worse than 1% defective. In broad terms, the standard called for a specific number of samples to be chosen based on the size of a known quantity of product. The number of defects in the sample determined if the product would be accepted or rejected; many plans allowed product to be shipped even if a certain number of defects were found in the sample.

By twenty-first century standards, it is incredible to believe that customers agreed to a plan that would provide them with product that was on average 1% defective, but nevertheless, that was the accepted standard for many years.

Mil 105 has been replaced by an American National Standards Institute (ANSI) counterpart, which unfortunately is still in use by some manufacturing facilities. The idea of agreeing to accept a certain level of defects from a supplier is counter to competitive manufacturing practices. And more importantly, the practice has led to the misconception that the percent defective in a random sample is an accurate indication of the level of defects in the product from which the sample was selected. The bead box demonstration helps to dispel this mistaken belief.

For the moment, assume the bead box, in fact, represents the output of a manufacturing process and the information in Table 4.1 is forwarded to a department manager each morning. The department manager might become somewhat concerned when he or she sees the results of Tuesday's output represented by Samples 4, 5, and 6. The manager's thought process might follow along the lines that on Monday all three shifts produced 2% defective product; on Tuesday, the same three operators produced product that was defective 6%, 8%, and 14%, respectively. When discussing the situation with the shift supervisors and operators they all agreed the heat and humidity on

Tuesday increased as the day wore on and became extreme on third shift. Therefore, the "excessive percent defective" on Tuesday was a result of heat and humidity.

At this time, I would like to remind the reader that the number of defects at the output of all 30 shifts recorded in Table 4.1 is a constant 5%, and the sampling results are a product of random chance. The really bad news is the manager, supervisors, and operators collectively created a new shop floor legend that heat and humidity cause yellow beads.

There are probably dozens of erroneous opinions and conclusions that could be formed about the weather, operators, equipment, raw material, and so forth, that caused the defect level to increase or decrease during the 2 weeks represented by the data in Table 4.1. However, weather and raw materials seem to be particularly popular scapegoats in manufacturing. This phenomenon of what appears to be an unstable attribute defect often leads to expensive activities. Monies and resources are expended to find the instability problem that does not exist. Barriers are created between people and departments (operators versus inspectors and production versus engineering) and unnecessary adjustments and modifications are executed—all for no good reason. The output from each of the 30 shifts is a constant 5% defective.

If the department manager had available the chart illustrated in Figure 4.1, the manager would know the process is stable and the centerline represents the constant percent defective under normal process conditions. Normal process conditions in this case would be no yellow beads are being added or removed from the box. Also, the manager would accept the fact that the variation in the 30 random samples is absolutely normal. Tuesday's sample showing a 14% defect rate is as normal to this process as the person at the casino dice tables who occasionally throws double ones, snake eyes, on the first roll.

CASE STUDY 4.1 PERCENT DEFECTIVE CABLE ASSEMBLIES

A manufacturer of injection-molded product was experiencing unacceptable levels of scrap and rework for a complex electronic component produced for the telecommunication industry.

A great deal of expense went into the manufacture of the product due to a labor-intensive assembly process that involved the manual connection of electrical components. Also, due to the nature of the finished product application, the customer required, and paid for, a 100% final inspection of electrical and visual (attribute) characteristics. The final inspection was reasonably effective on the basis that there were few customer returns or complaints regarding quality. However, management was extremely concerned with the level of visual rejects found at final inspection. The customer was reasonably satisfied, but the product line was losing money due to in-house rejections, scrap, and rework primarily due to visual defects. Final inspection reports generated on a daily basis indicated the reject rate on some days exceeded 25% over the three-shift operation. On other days, one shift would have zero rejects and the very next shift would generate over 15% rejected assemblies. Engineering efforts and meetings with shop floor associates seemed to result in short-lived improvements, but the higher reject rates always came back. There were numerous disagreements regarding the causes of the rejects, but everyone agreed that the vast majority of rejects were due to visual defects and that there was neither rhyme nor reason to the reject patterns.

The accounting department had factored in a 5% reject rate when the full cost of manufacturing the product had been calculated. Out of frustration with the excessive reject rate, management had required a reason for excessive rejects be identified any day the visual reject rate exceeded 5%. A meeting of the vice president of operations and the production staff was held at 10:00 a.m. each of the days after the reject rate had exceeded 5%. On most mornings, the production staff had little or nothing to offer as to the cause of the unacceptable reject rate.

Perhaps the production staff could not find a viable reason for a defect rate of 5% because the normal variation for attribute percent defect rate exceeded 5%. Perhaps it would be perfectly normal to experience a 15% attribute defect rate on one day just as it would be perfectly normal to get 14 yellow beads out of a 100-bead random sample during a bead box demonstration.

It is important to recognize that even though 100% visual inspection was being performed on every lot, that each lot, in fact, represents a random sample selected from a continuous stream of product that is perhaps producing a constant level of defects during periods of stability. The number of rejects in any one lot is a result of that lot being a random sample of the entire process population.

Unfortunately, this manufacturing company was getting blindsided by the bead box effect.

With respect to the excessive number of rejects, the question was finally asked—"Is the unacceptable level of rejects due to process instability or process incapability?" The answer was not forthcoming because the reject data at final inspection had never been analyzed in terms of "normal" (binomial) variation of percent of attribute defects (rejected product).

It was decided to perform a process capability study and construct a P chart to determine if the wide-ranging reject rate being experienced was due to instability or due to a very wide stable distribution of percent defect data.

Management expressed a strong desire for quick resolution of the problem due to the excessive cost being incurred on a daily basis. In view of management's concern, the suggestion was made that historical data from final inspection, properly analyzed, might very well provide insight into the nature of the problem.

Historical reject data in the form of percent defective for attributes is often available within manufacturing facilities, but the data are seldom used for problem solving. A great deal of effort is expended when inspecting product and usually at considerable indirect labor cost, but very little benefit to the process is derived from the effort put forth.

In this case, the final inspection records were maintained primarily in order to generate the daily reject report. The percent defective data were recorded for each operator on each shift; however, the data were not categorized by defect type, which is not compatible with normal use of such data for problem-solving activity. The ideal situation would have required the rejects to be quantified by type, which would allow efforts to be concentrated on one of the major categories. It is not unusual for a manufacturing facility to combine multiple types of defects into one number for reporting purposes, but this leads to various opinions as to which type of defect is the most serious, and efforts to reduce the reject rate are subsequently diluted.

The decision was made to review the final inspection data from the previous 30 days from workstation 1 in order to determine if the process was stable with respect to attribute defects. It was further decided to eliminate variation due to possible operator technique differences in the manual assembly of the product by using data from one assembler normally assigned to workstation 1. Analyzing historical data is not the best way to perform a process capability study, but conditions at this company were such that expediency trumped statistical purity, and to quote the father of statistical process control:

> The fact that the criterion which we happen to use has a fine ancestry of highbrow statistical theorem does not justify its use. Such justification must come from empirical evidence that it works.
>
> **W.A. Shewhart**

Past experience with using historical data to jump-start a problem-solving journey provided a level of confidence that this timely and inexpensive effort would provide the problem solvers with some direction and management with some sorely needed viable answers.

Table 4.2 represents the percent defective for the previous 30 days recorded for the first-shift assembler working at station 1.

Figure 4.2 is the P chart created with the Table 4.2 data. The centerline of the chart is 6%, and just as in the bead box example, the 6% represents the average percent defective for the assembler and workstation combination

Table 4.2 Percent Defective Historical Data for Assembler 1 Workstation 1

Shift	*Lot Size*	*Percent (%) Defective*	*Shift*	*Lot Size*	*Percent (%) Defective*
1	140	12	1	146	5
2	145	5	2	155	9
3	104	6	3	178	12
1	167	6	1	180	1
2	114	0	2	154	12
3	140	2	3	154	1
1	158	0	1	178	12
2	122	14	2	157	2
3	101	2	3	178	8
1	130	0	1	143	0
2	122	10	2	176	2
3	164	0	3	148	13
1	122	17	1	178	0
2	112	0	2	176	6
3	150	6	3	178	14

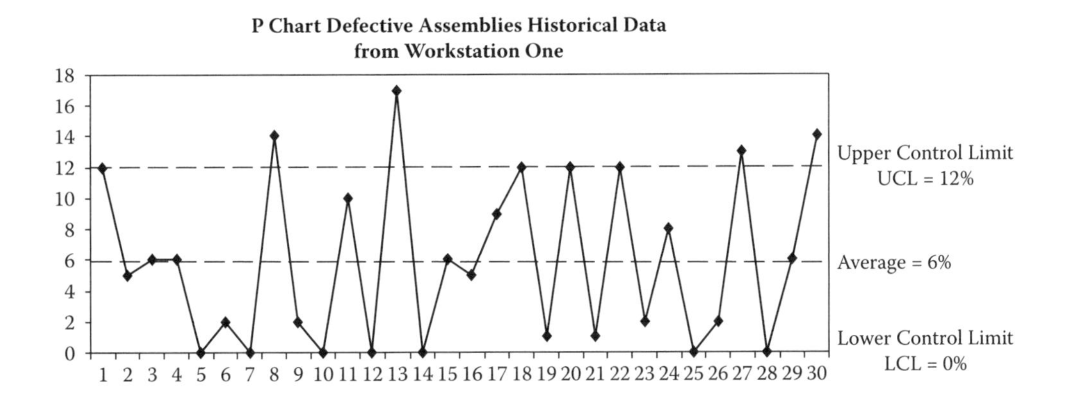

Figure 4.2 P Chart defective assemblies historical data from workstation 1.

that was being studied. The UCL is 12%, and the lower control limit is zero; the data points appearing above the UCL are indications of instability.

The problem appeared to be one of instability. When a process has been identified as being unstable, the very first responsibility is to identify and eliminate the root cause of the instability.

A review of the data and the control chart with a group of managers, molding technicians, engineers, final inspectors, and the three first-shift assemblers was reinforced with a brief bead box demonstration. The group agreed the control chart indicated the process was unstable with respect to attribute defects.

A general discussion ensued regarding the types of defects the inspectors rejected and which defects might be a result of the assembly process. The final inspectors offered what they thought were the three most serious categories of defects that were, in descending order, exposed connectors, voids, and streaks. The inspectors also agreed the exposed connector defect seemed to come and go, which would fit the instability profile indicated by the P chart.

When using historical data, the procedure outlined in the flow diagram of Figure 3.6 obviously has to be applied after the fact. The meeting with the group provided ample reason to conduct a walk around of the process looking for possible causes of instability that might result in exposed connectors. The review began at the workstations where the connectors were assembled and insulated. A lot was selected at random and followed through the process steps subsequent to assembly. As the lot moved through the remainder of the process steps, a number of the assemblies began to exhibit exposed connectors, indicating the insulating process was not very robust.

The review moved back to the assembly workstation in order to observe the process more closely. One of the final steps in the assembly process was to slip a length of shrink tubing over a wire lead before soldering the wire lead to a connector; when all the wires were soldered in place, the shrink tubing was slipped over the individual soldered connections and a heat gun was applied to the shrink tubing which served to electrically insulate each connector. An engineer pointed out the operators were cutting the shrink tubing by hand, resulting in different cut lengths and angled cuts. As the review moved from one workstation to another, it was also noted that different operators applied the heat gun using various techniques. It was determined the variation in the cut and angled lengths combined with different operator techniques of applying heat to the shrink tubing quite probably were contributing to the defect "exposed connectors."

Table 4.3 Percent Defective after Installation of Fixtures Assembler One/Workstation One

Lot Size	*Percent (%) Defective*	*Lot Size*	*Percent (%) Defective*
130	5	126	6
124	5	143	2
137	2	124	3
126	3	124	3
163	5	124	4
141	0	97	4
124	1	160	3
99	3	157	0
109	4	108	4
153	6	151	5
117	4	122	5
96	6	104	0
117	3	142	3
117	3	132	5
146	5	158	4

A fixture for cutting the shrink-tubing square and to the required length plus a fixture to secure the assembly during heat application were installed on all four workstations.

Table 4.3 shows the first-shift operator's total defect experience for the 30 days following the installation of the fixtures. The percent defective for exposed connectors, separately identified by the final inspectors, was negligible, clearly indicating the addition of the fixtures served to dramatically reduce the overall percent defective due to exposed connectors. This improvement is graphically represented in Figure 4.3, which represents the P chart from Figure 4.2 with the new data added.

The new data were used to construct the P chart found in Figure 4.4. This chart indicates the process with respect to total percent defectives is now stable with a lower overall average percent defect rate of 3.5% and a UCL of 8%.

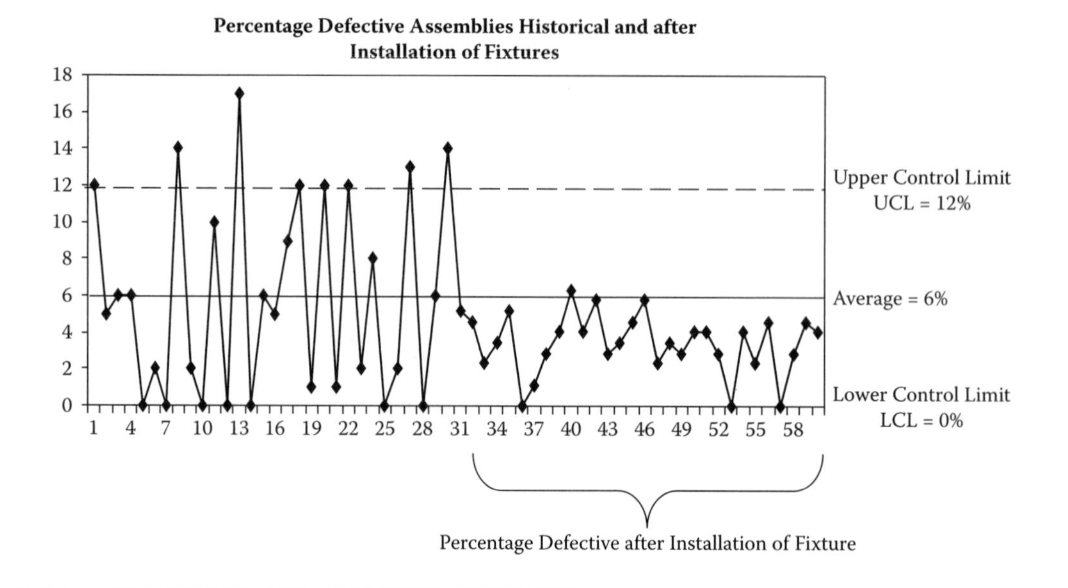

Figure 4.3 Percent defective assemblies, historical and after installation of fixtures.

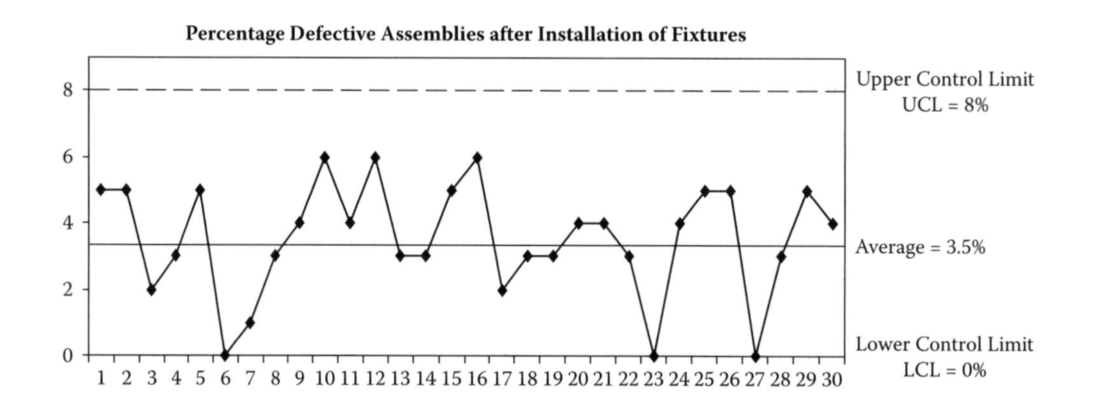

Figure 4.4 Percent defective assemblies after installation of fixtures.

Current data were analyzed from the other operators with almost identical results. The 12 process streams were now stable with respect to all types of attribute defects, including the much-reduced "exposed connector" defect, with a lower average and a lower UCL. Going forward, management may expect the average percent defective for all categories of defect to be 3.5%, with the daily defect rate for each assembler to vary randomly about the average with an occasional data point to be at or near the 8% upper limit. Management now had a tool that would enable them to easily stay in touch with the overall percent defective for this process and to make more informed inquiries of the production department with respect to attribute defects stemming from this product line.

After the Process Capability Study Is Completed

A P chart for all categories of defects can be a very useful management tool, but the highest and best use of the P chart concept is realized when it is applied to a single defect type for problem-solving purposes. Therefore, in order to facilitate future investigations of attribute defects, the final inspectors were directed to begin categorizing the rejects by type.

At the end of a month, a Pareto chart of attribute defects by type was reviewed, and a P chart was created for the next most serious defect type that turned out, as the final inspectors had estimated, to be voids.

A P chart for voids was constructed based on the final inspection data and maintained by the first-shift supervisor. Each morning, the percent defective data for "voids" from the previous day were recorded on the supervisor's P chart, and the supervisor's responsibility was to alert the molding technicians to any periods of instability. For instance, if the daily percent defective on the Voids P chart exceeded the UCL, the supervisor and the molding technicians would attempt to identify the cause. Whenever the cause of instability was identified, a note was made on the P chart. If the cause could not be identified, the engineering department was involved.

This problem-solving effort could proceed in a number of directions depending on the combined process knowledge of the molding technicians, supervisors, and engineers:

- A comprehensive review of all the P chart notations could be conducted to determine if there was a recurring cause that required attention.
- A walk around with engineering and supervisors could be conducted checking out standard procedures, equipment, and raw material related to the molding process.
- A Pareto chart could be developed for each of the injection-molding presses to identify if there were one or more presses generating the majority of voids.
- A simple designed experiment (discussed in Chapter 5) could be conducted to identify the process parameter causing the increase in voids.

How the problem solvers proceed from their review of the P chart is not as important as the fact they will proceed using proven tools and techniques as opposed to drawing conclusions based on anecdotes, opinions, and forceful personalities.

They should not fall into the trap of subordinating the attribute problem, as many companies do, by entering into negotiations with the customer as to exactly how big a void can be accepted.

These negotiated settlements with customers often end in written definitions of "when a streak is not a streak" or "when a void is too big or too many" that they rival some of the ancient metaphysical arguments of how many angels can dance on the head of a pin.

The basic definition of an attribute defect is it is there or it is not. Settling on definitions of "acceptable voids" will increase inspection and rework costs, build barriers between inspectors and manufacturing associates, and leave a bad impression on the customer. It would be so much more productive to assign resources to identify and eliminate the cause of the attribute defect rather than attempt to get the customer to accept minor streaking or no more than X number of voids no larger than Y in an area no greater than Z.

The preceding study performed using historical data and total percent defective data is not the classic method of performing an attribute process capability study, but economy and common sense must trump standard protocol in the competitive environment in which we all operate. It was assumed the defect "exposed connectors" was the most serious category, and it was assumed the historical data were not contaminated by process changes or inexperienced inspectors in the final inspection department. The results indicate these along with other assumptions were valid, because the study significantly improved the process. And that is the goal of the process capability study.

Meaningful Attribute Capability Concept

Capability in terms of the capability index (Cpk) for a variable characteristic, discussed in detail in Chapter 3, depends on comparing the amount of characteristic variation to the customer specification. The arithmetic for variable Cpk cannot be applied to attribute data because customers do not typically offer a specification for attribute data in terms of 5% defective ±2%.

With the exception of an AQL agreement between customer and supplier, I have never seen a customer specification for the number of allowable attribute defects. And as stated earlier, the AQL philosophy leads to certain misconceptions regarding the actual percent defective in a specific lot of product. However, it is common for manufacturers to incorporate an anticipated percent of scrap into the costing formula of a product. This anticipated cost of scrap is often spoken of in terms of an allowable scrap level, and I contend this allowable scrap level, determined by management, can be utilized as an *internal* Cpk goal.

All things considered, I adopted a unique approach to dealing with attribute Cpk which is a more practical application of the concept than another method used by some facilities which involves Z statistics and provides an index in terms of defective parts per million. The approach I recommend takes into account the internally established allowable scrap level and, for this reason, is generally more meaningful to management than the aforementioned method.

In Case Study 4.1, discussions with the accounting department led to an understanding that a 5% allowable scrap factor was included in the costing formula for the product in question. The process capability study performed after the installation of the fixtures indicated overall attribute percent defective was stable with a defect average of 3.5% and an UCL of 8%. If we were to consider the internally established allowable scrap level of 5% to represent an "upper specification," we would have to declare the process as internally incapable, because a certain number of shift outputs will be as high as 8% defective.

If we compare the allowable scrap rate of 5% to the average percent defective (P Bar) of 3.5% and the P chart UCL of 8%, we can develop a relative Cpk metric.

The arithmetic for this suggested method is as follows:

$$\text{Cpk} = \frac{(\text{Allowable Scrap Level} - \text{P Bar})}{(\text{UCL} - \text{P Bar})}$$

$$\text{Cpk} = (5\% - 3.5) \div (8\% - 3.5\%)$$

$$\text{Cpk} = 1.5 \div 4.5$$

$$\text{Cpk} = 0.333$$

The Cpk of 0.333 clearly indicates the process is presently incapable of achieving the internal goal of 5%. Now management has an initial, meaningful metric against which progress can be measured. As the problem-solving activity goes forward using techniques such as were outlined above, management should expect the Cpk to increase over time. When P Bar reaches 1.2% for all categories of attribute defects related to this product line, the process would be deemed capable of meeting the internal requirement.

When P Bar reaches 1.2%,

$$\text{Cpk} = \frac{(\text{Allowable Scrap Level} - \text{P Bar})}{(\text{UCL} - \text{P Bar})}$$

$$\text{Cpk} = (5\% - 1.2\%) \div (2.876)$$

$$\text{Cpk} = 3.8 \div 2.876$$

$$\text{Cpk} = 1.32$$

Of course, achieving a Cpk of 1.32 would not mean the end to improving the process in a truly competitive organization.

Once again, this method of calculating attribute Cpk has no relationship to the defective parts per million components that are the basis for an attribute Cpk calculation suggested by some. This method is suggested only for internal use by management as a number that will indicate improvement and a reduction of defects toward the ultimate goal of zero.

Chapter 5

Addressing Instability and Incapability

If you are out to describe the truth, leave elegance to the tailor.

Albert Einstein

Introduction

Performing a process capability study requires discipline, and when performed properly the benefits can be manifold. This chapter will address a suggested protocol for dealing with instability and incapability if they are recognized as results of the study.

- In Case Study 2.1, the molding process was discovered to be stable but incapable due to a change in *raw material.* When the customer agreed to change the blueprint specification, the belt width became capable.
- In Case Study 3.1, the insulation wall thickness was determined to be stable but technically incapable due to *operator techniques* that placed the process average too close to the upper specification. This was not in the best interest of the company, and when the process average was adjusted, the insulation wall thickness could then be described as capable.

- In Case Study 3.2, the crimp diameter was found to be unstable due to an unstable *measurement process* and *operator technique* differences in setting feed roll tension. When the instability was removed by modifying the measurement process and the installation of a constant tensioning device, the crimp diameter process became stable and capable.
- In Case Study 4.1, the cumulative cable percent defect was unstable. When the *operator technique* was standardized by the installation of fixtures, the significant defect "exposed connectors" was removed, and the process was stable with respect to attribute percent defective. The cumulative percent defective was determined to be incapable with respect to the internal allowable percent defective.

The studies listed above are generally representative of the process capability studies I have been involved in over the past 30-odd years.

- In case studies 2.1 and 3.1, the lack of capability was the problem.
- In case studies 3.2 and 4.1, the lack of stability was the problem.

Instability

In Case Study 3.2, "Crimped Terminal Diameter Customer Returns," the instability was due to an inadequate measurement process and operator technique. The inadequate measurement process was corrected before the process capability study was performed as a result of following the flow diagram of Figure 3.6. The operator technique contribution to the instability was a result of discussions with the operators after the study. These discussions were conducted with the knowledge gained by the study that the problem was one of instability.

In Case Study 4.1, "Percent Defective Cable Assemblies," the instability was due to an inadequate standard operating procedure (SOP) and lack of appropriate fixtures. After the study, armed with the knowledge that the problem was one of instability, observation of the process led to the installation of fixtures.

Of course, all forms of process instability are not as easy to identify as those discussed in these two case studies, and not all forms of instability can be attributed to inadequate measurement processes, operator technique differences, or inadequate SOPs. The root cause of some forms of instability can be the result of raw material, equipment, or local environment.

Figure 5.1 is a fishbone diagram that illustrates the six elements of every manufacturing process which need to be considered when addressing a process problem of instability or, for that matter, incapability.

I would like to take this opportunity to remind the reader that the term *instability*, when applied to a manufacturing process, implies the average performance of the process output has unexpectedly shifted. (See Figure 5.2.)

If the analysis of the product control chart resulting from a process capability study indicates instability (violation of at least one of the instability detection rules), the investigation should begin with a review of the six elements of the process.

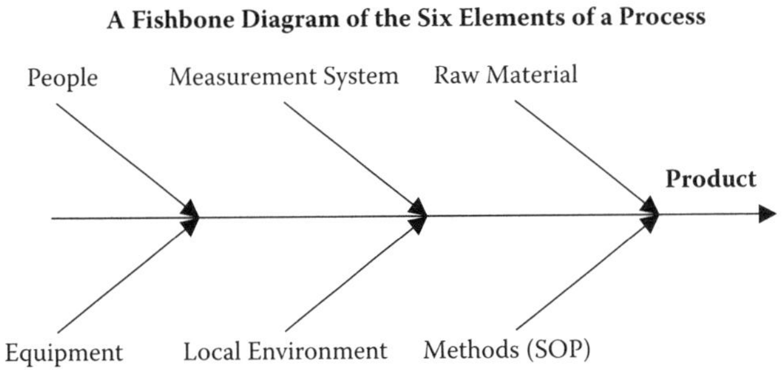

A Fishbone Diagram of the Six Elements of a Process

People Measurement System Raw Material

Product

Equipment Local Environment Methods (SOP)

Figure 5.1 A fishbone diagram of the six elements of a process.

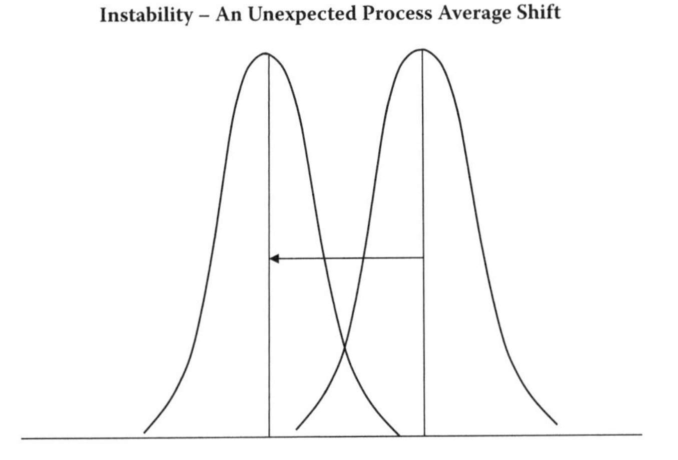

Figure 5.2 Instability—an unexpected process average shift.

Process Checklist

- Measurement process—The measurement process is qualified before the process capability study is performed; therefore, it can usually be discounted as a root cause of the problem.
- Methods—Also in preparation for the process capability study, a review of the SOP is made to ensure the process parameters are set up accordingly. However, it would be advisable, as in the exposed cable connector study, to review the SOP after the fact knowing now that we are looking for a cause of instability.
- People/operator technique—A full review of the process capability study results with the operators is highly recommended. The operators will feel they are part of any solutions that are developed, and as in the crimp diameter study, subtle operator technique differences are often uncovered.
- Local environment—An open window, a fan directed at an operator's workstation that is near a thermocouple, an AC vent output that is redirected by means of a piece of contraband cardboard, and so forth, are examples of local environment that can cause instability. Once again, this type of information can be revealed during a meeting with the operators.
- Raw material—A frequent cause of instability can result from differences in raw material characteristics which vary from batch to batch and from supplier to supplier. Sometimes raw materials suppliers will change their process, which can adversely affect the customer's process. A problem of instability due to raw material typically requires the use of a tool that will be discussed shortly.
- Equipment—Obvious equipment anomalies would be noticed during the walk around before the process capability study begins. Of course, various equipment parameters such as line speed, tension, cycle time, and dwell time, need to be considered when attempting to identify the cause of instability.

Problem with Random Equipment Adjustments

When the cause of the instability, identified by a violation of one of the instability detection rules, cannot be identified by reviewing the usual suspects, such as operator technique or SOP, a typical group of problem solvers will often begin to make adjustments to equipment parameters. Opinions as to the effect various equipment parameters such as line speed, heat, and tension, have on the process are usually not in short supply. Making adjustments to these equipment parameters is an attempt to compensate for an obvious process average shift. A universally popular practice in industry when attempting to eliminate instability is to make only minor adjustments to one process parameter at a time. This practice stems from the general belief that

- If only minor adjustments are made, there is less risk of making a bad situation worse. Shop floor problem solvers are hesitant to make more bad product when attempting to solve a problem while simultaneously manufacturing product.
- If several parameters are adjusted at the same time and a positive result is achieved, the group will not know which of the several adjusted parameters caused the improvement.

There are two fallacies associated with these practices:

- In making a minor adjustment to an equipment parameter that will, in fact, have a positive effect on the process, the improvement might be so small it would not necessarily be recognized. Figure 5.3 illustrates this concept.
- The practice of adjusting only one equipment parameter at a time ignores the potential that the product defect is a result of an *interaction* between two or more equipment parameters.

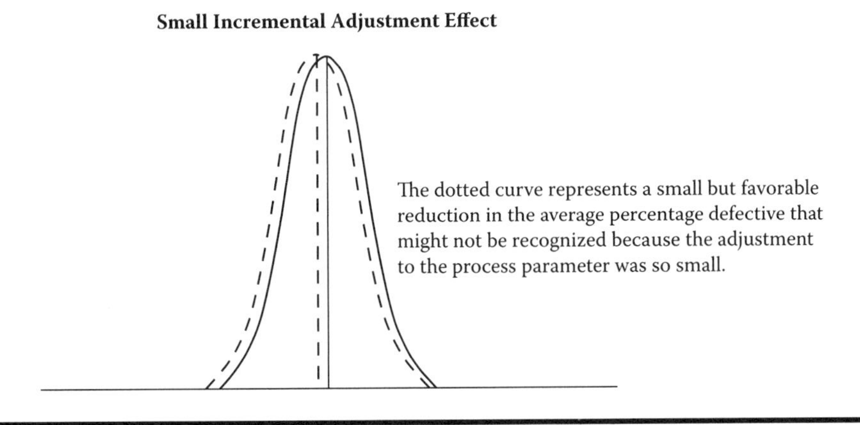

Figure 5.3 Small incremental adjustment effect.

Interaction

The dictionary defines *interaction* as "A reciprocal, effect or influence." (http://dictionary.com)

A familiar interaction is the wind chill factor reported on the Weather Channel. A person feels a certain effect on his or her exposed body parts when the temperature is at 20°F, with no wind. When a 35-mile-an-hour wind accompanies the same 20°F temperature, the effect felt on unprotected skin is very different. There is an interaction between temperature and wind.

In industry, a simple interaction exists when the effect on the process output when one process parameter is at a certain level depends on the level of a second process parameter. Many chronic process problems exist in industry today, because important interactions have not been discovered as a result of changing only one process parameter at a time.

Making random adjustments to equipment for the purpose of eliminating instability is expensive, ineffective, and inefficient. One alternative that works very well is a simple designed experiment.

CASE STUDY 5.1 MISSING CLIPS

A manufacturer of assembled metal components was experiencing a serious problem of missing clips on finished product, and a team had been assigned to identify and eliminate the root cause of the missing clips. The team conducted an informal brainstorming session and concluded there were only a handful of process parameters that might cause missing clips.

The process parameters that might cause missing clips were determined to be

Clipping station
Air pressure
Speed of clipping
Steel thickness
Steel width
Operator
Clipping gun

It was recognized that operators and supervisors favored making adjustments to air pressure and speed of clipping when the percent of missing clips began to run higher than "normal." In view of this fact, air pressure was the first parameter the team decided to manipulate to see if they could get a positive result.

The air pressure for the gun mounted on clipping station 1 was set according to the SOP, and during the next hour a 50-piece sample of the product was selected and checked for missing clips. The random sample of 50 contained 6 units with missing clips for a 12% defect rate. Then the air pressure was raised 1 PSI above the SOP required setting and another 50 units were selected over the next hour. The 50 units were inspected, and 8 units were reported to be defective for a 16% defect rate. This was interpreted as the air pressure had been adjusted in the wrong direction. At this point in time, the air pressure was decreased to 1 PSI below the SOP. The 50-piece sample contained only 1 defective unit with missing clips for a mere 2% defect rate. It was late in the day, and the team believed they had experienced a breakthrough. An entry was made in the supervisor's log book that all clipping stations were to have the air pressure set at 1 PSI below SOP. The team departed fully anticipating the second and third shifts would have a much reduced percent of defects due to missing clips.

The next morning, the team was greeted with the bad news that all of the clipping stations were experiencing the "normal" number of missing clips once again.

The team reconvened and decided to make further adjustments to the air pressure on several of the clipping stations. But after a morning of trials, they could not repeat their short-lived success of the previous day.

I would like to remind the reader that the normal percent defective of yellow beads in the bead box would be the average of 5% with a lower range of 0% and an upper range of 14%. We know this to be true because a P chart was created using the results of 30 random samples. If we think of the clipping process in terms of the bead box, it *might* be perfectly normal to experience 12% defectives in one sample of 50 components and then experience a 16% defect rate in the very next sample followed by a 2% defect rate. It was just random chance the team found only 2% defective product when the air pressure was set at 1 PSI below SOP.

The team did not realize it, but they had been victims of the bead box effect.

It was finally decided that air pressure was not causing the problem, and attention was brought to bear on the adjustable parameter speed of clipping.

The speed of the clipping process was increased by five clips per minute above SOP requirement, and the output was monitored. Then the speed was reduced by five clips per minute below the SOP. For a brief period when speed was reduced, "missing" clips were at a lower level than "normal," but that did not last long.

After 3 months, the team had trialed each of the seven chosen parameters in the same manner as they had trialed air pressure and speed with no permanent success. That is, the team made small, incremental adjustments to a single process parameter and drew conclusions regarding the overall percent defective based on a random sample. The team was not considering possible interactions by adjusting more than one parameter at a time; they were not considering that a minor adjustment might cause a change that would not be detectable. And, of course, the evaluation of resultant percent defective data was being made without any knowledge of the type of problem they were dealing with—they had no idea if the problem was one of instability or of incapability.

At this point, the team was encouraged to take a step back and perform a process capability study in order to create a percent defective chart (P chart). The purpose of this suggestion was to

Determine if the process was stable with respect to missing clips.

Enable the team to identify the actual normal range of percent defective for missing clips.

Several members of the team, including the team leader, had very positive past experiences with process capability studies, and yet they went 3 months without using the technique. There seems to be a tendency in industry to try the informal quick fix first and then, after all else has failed, go to the toolbox of proven techniques.

This is why every problem-solving effort must begin with the prime question: "What kind of problem is it, is it one of instability or is it one of incapability?" The answer cannot be offered if a process capability study has not been performed.

A process capability study was agreed on, and the flow diagram of Figure 3.6 was closely followed. Due to the nature of the inspection, a measurement study was not necessary.

Samples were selected every half hour for the next three shifts. See Figure 5.4 for the resultant P chart.

Once the P chart was created, it was obvious the process was producing, on average, 12% defective with a lower limit of 0% and an upper limit of 26%, and the process was unstable. Note on Figure 5.4 the run below the centerline beginning at data point 5 and the run above the centerline beginning at data point 31. Also, two data points fall outside the upper control limit.

It was then suggested the team meet with the operators and use the Figure 3.6 flow diagram as a guide to check out the six elements of the process. Another equipment walk around, SOP check, and a second-operator

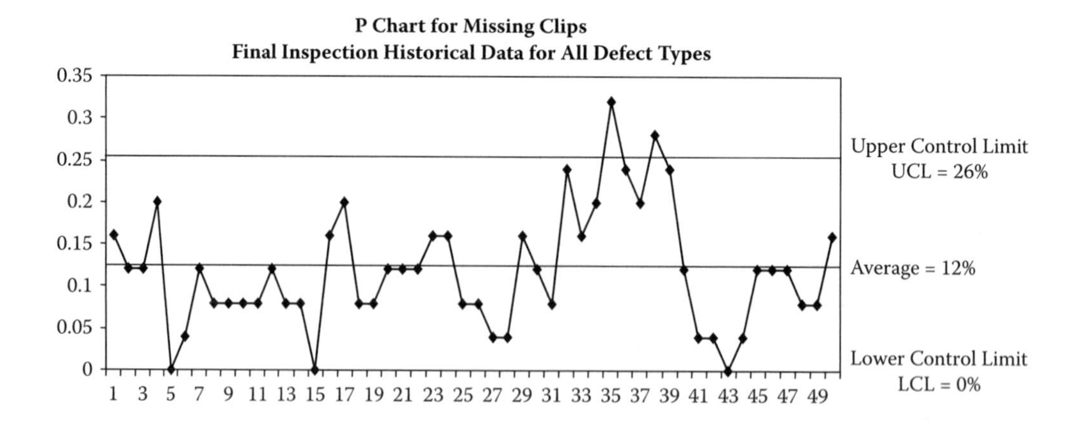

Figure 5.4 P Chart for missing clips, final inspection historical data for all defect types.

interview revealed nothing of interest leaving raw material quality and equipment parameters open to investigation.

The nature of the two process elements raw material and equipment parameters were well suited as subjects for a *very simple* designed experiment—an experiment that could be designed to include the possibility of an interaction between process parameters that might be contributing to the missing clip problem.

I emphasize the term *very simple* because design of experiment (DOE) technology is an entire complex study unto itself. However, there are several very easy to understand techniques that can be used on the shop floor without any of the complex mathematics or elaborate software that usually accompany the design, execution, and analysis of a designed experiment.

The group was asked to choose the three parameters from their brainstorming session that they considered to be the most likely causes of the missing clip problem. They chose

Air pressure
Speed
Raw material (steel) thickness

The team was then asked to choose a high level and a low level for each of the process parameters, and to be bold in the selection of the levels. High level was designated as a plus (+), and low level was designated as a minus (–).

As mentioned earlier in the chapter, when attempting to solve a problem on the shop floor by making adjustments to process parameters, people tend to make minor adjustments. In making minor adjustments, a minor improvement might result, but it would be so small that it would not necessarily be recognized. Figure 5.3 illustrates how a minor adjustment might not be recognized as important.

If the three chosen process parameters are to be properly evaluated for their effect on missing clips, it will be necessary to choose levels that will drive the process to create a significant difference in the percent of missing clips. In order to recognize a significant change to the percent defective, it would be necessary to choose levels for the three selected parameters so they are relatively low and high (bold enough) with regard to the SOP requirements.

For instance, if the SOP calls for the air pressure to be set at 15 PSI, it might be a good idea to choose a low level of 10 PSI and a high level of 20 PSI. Might one of these settings create a high level of missing clips? Yes. And that would be good news, because that would indicate that air pressure is a dominant process parameter that contributes to the condition "missing clips."

The objective of a designed experiment is not to make good product. The objective is to identify the dominant process parameters with respect to the defect being investigated.

A word of caution is in order. When choosing the levels for an experiment such as is under discussion, it is a good idea to have engineers involved. Although we wish to be bold in selecting the high (+) and low (–) levels, we do not want to damage the equipment or otherwise engage in unsafe practices. But we do expect with the appropriate high (+) and low (–) levels that some defective product will result.

The air pressure and clipping speed parameters could be easily adjusted to their respective high (+) and low (–) levels to satisfy the requirements of the experiment. The group had to be a little creative to adjust the high (+) and low (–) levels of material thickness. Inspectors went to the warehouse and sorted through the inventory of rolled steel until they found one roll of steel at the low end of the thickness specification and one roll that was just above the upper specification for thickness.

Once the parameters and their respective high and low levels have been chosen, a series of eight production runs (experiments) are conducted according to the levels found in Table 5.1. Consider, for example, the first three runs:

- Run 1 would require steel thickness, air pressure, and speed set at the high (+) levels chosen by the team. A number of units would be processed under these conditions, inspected, and the percent defective recorded in the Response column.
- Run 2 would require steel thickness and air pressure set at the high (+) levels, speed set at the low (–) level, and the percent defective units produced would be recorded in the Response column.
- Run 3 would require steel thickness set at the high (+) level and air pressure and speed set at the low (–) levels, etc.

To expedite the experiment and minimize interference with production, the raw material parameter was placed in the first column so that the roll of steel would only need to be changed once. Air pressure and speed, as mentioned above, were easily adjusted.

When the high level (+) of raw material was installed and air pressure and speed were set at their high (+) levels, the experiment was begun. The eight runs were conducted according to the levels recorded in Table 5.1. Each member of the team had a copy of Table 5.1 to ensure the levels were set correctly for each of the eight runs.

The reader will notice that this table provides for all combinations of levels to be run for all three process parameters. In design of experiment terminology, process parameters are referred to as *factors* and because this experiment allows for a "full" combination of two levels (+ and –) for all three *factors*, this particular type of experiment is called a *three factor, full factorial*—a fanciful term for a very simple tool.

The team was instructed to produce 50 units for each of the eight runs, inspect the components for missing clips, and record the percentage of units with missing clips in the Response column on a copy of Table 5.1.

Table 5.1 Three-Factor, Two-Level Experimental Matrix

Run	*Material Thickness*	*Air Pressure*	*Speed*	*Response*
1	+	+	+	
2	+	+	–	
3	+	–	–	
4	+	–	+	
5	–	–	–	
6	–	–	+	
7	–	+	+	
8	–	+	–	

Table 5.2 Three-Factor, Two-Level Experimental Matrix

Run	*Material Thickness*	*Air Pressure*	*Speed*	*Response (%)*
1	+	+	+	4
2	+	+	–	8
3	+	–	–	8
4	+	–	+	10
5	–	–	–	2
6	–	–	+	4
7	–	+	+	8
8	–	+	–	10

Note: Fifty units produced for each of eight experiments, and the percent of defective units recorded in the Response column.

Table 5.2 illustrates the eight runs with the responses for each run filled in with the percent defective.

At the conclusion of the eight runs, the first comment from the team members was related to the differences between the various responses—from 2% to 10%. When the results of an experiment produce a series of responses ranging from very low to very high, there is a confidence that at least one of the chosen parameters is having an effect on the process.

Now the team was instructed to rank the responses in ascending order, making certain that each response retained its unique combination of pluses and minuses. This is a quick and nonmathematical way to determine if there were any patterns of low and high levels in conjunction with lower percentages of missing clips.

In Table 5.3, the Response column shows the responses ranked in ascending order alongside the corresponding levels that created the responses. It was clear to the team that the lower percentages were from those runs where both material thickness and air pressure were at their low levels.

In addition to material thickness and air pressure being important to the percentage of defective components due to missing clips, there appeared to be an interaction between the two parameters:

When both parameters are low, the percentage of missing clips is low: 2% and 4%.

When one parameter is high and the other is low, the percentage of missing clips is high: 8%, 8%, 10%, and 10%.

When both are high, the percent of missing clips is mixed: 4% and 8%.

A suspected interaction can be confirmed with a little bit of arithmetic.

Note that any two columns in Table 5.3 contain two rows with a – – combination, two rows with a + + combination, two rows with a + – combination, and two rows with a – + combination. The interaction of material thickness and air pressure is determined by averaging the responses of each of the combinations of pluses and minuses found in the raw material and air pressure columns in Table 5.3.

Refer to Table 5.4 for the simple arithmetic used to determine the interaction between material thickness and air pressure.

Table 5.3 Three-Factor, Two-Level Experimental Matrix with Responses in Ascending Order

Run	*Material Thickness*	*Air Pressure*	*Speed*	*Response (%)*
5	–	–	–	2
6	–	–	+	4
1	+	+	+	4
2	+	+	–	8
3	+	–	–	8
7	–	+	+	8
4	+	–	+	10
8	–	+	–	10

Table 5.4 Plotting an Interaction

	Material Thickness	*Air Pressure*	*Average of Two Runs with Identical Signs*
2% + 4%/2 = 3%	–	–	3%
8% + 10%/2 = 9%	–	+	9%
4% + 8%/2 = 6%	+	+	6%
10% + 8%/2 = 9%	+	–	9%

Notes: Material thickness low, air pressure low: low percent defective.
Material thickness low, air pressure high: high percent defective.

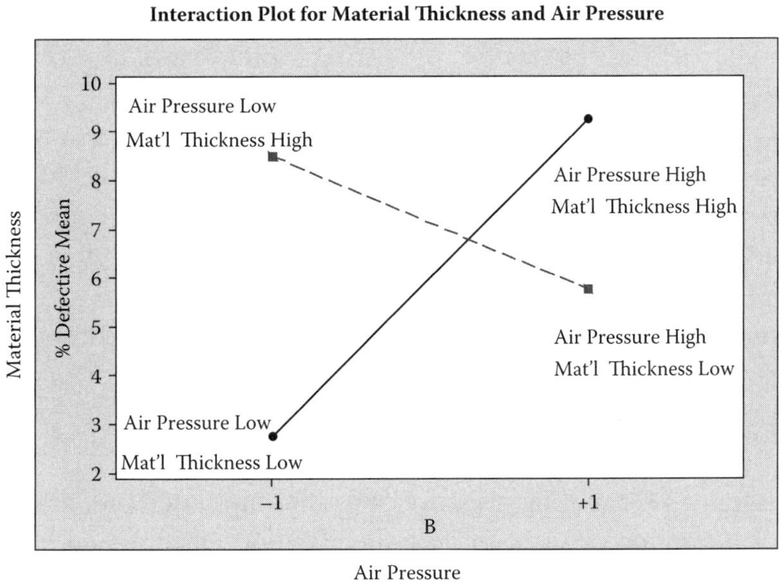

Figure 5.5 Interaction plot for material thickness and air pressure.

Graph paper can be used to plot this interaction. The graphical representation of the interaction between raw material and air pressure can be found in Figure 5.5.

It became obvious that, to a certain extent, material thickness and air pressure interacting had contributed to the instability.

If an operator had the air pressure set to some low level and the rolled steel being used was on the low side of thickness, the missing clip percent defective would be very low. If the next roll of raw material was on the high end of thickness, the missing clips percent defective would increase dramatically. If adjustments to air pressure were made, an increase in air pressure would serve to place the percent defective somewhere between pretty good and very bad. At some point with the air pressure now at some higher level, raw material of a lower thickness would be introduced and the percent of missing clips would go down somewhat but not to the lowest level that could be achieved if air pressure were to be reduced from its high level.

It was determined that modifications would have to be made to the raw material or the clipping guns in order for the process to be improved. For the interim, only rolls of raw material at the low end of the thickness specification would be used.

Need for Science

The dictionary defines *science* as "Reflecting a precise application of facts or principles; proficiency." (http://dictionary.com)

Once a problem has been identified as instability, the root cause can be determined by following a discipline. The discipline begins with reviewing the six process elements described in Figure 5.1. If the root cause does not become obvious by means of observation and interviews with shop floor associates, then science must be applied.

Too often in industry, important decisions are made based on opinions as opposed to facts. Opinions about processes are formed over the years as a result of individual experiences:

- An operator lowered air pressure at a time the raw material happened to be running on the low end of specification and missing clips were few. And she formed the opinion that low air pressure reduces the number of missing clips.
- Another operator raised air pressure at a time the raw material was running on the high end of thickness specification and had fewer missing clips. He formed the opinion that high air pressure reduces missing clips.

Both operators were right, and both operators were wrong, and neither operator had the benefit of the fact that an interaction existed between raw material and air pressure—an interaction that can only be determined by using science. In this case, the science is the designed experiment.

Analysis of Variance (ANOVA)

Another prevalent practice in industry when attempting to solve problems of instability involves collecting data before and after a process change is implemented, and then comparing the averages of the two data sets in order to make a decision based solely on judgment.

Figure 5.6 illustrates two very different normal curves. Curve A is significantly wider than curve B. Curve A has two data sets x_1 and x_2. Data set x_1 represents the average of 25 measurements taken before a process change was implemented. Data set x_2 represents the average of data collected after the process change was implemented.

Curve B_1 contains data set x_3, the average of 25 data points selected before a process change was made. And curve B_2 contains data set x_4, which is the average of 25 data measurements collected after the change was implemented.

Please note that data sets x_1 and x_3 are numerically identical, and x_2 and x_4 are also numerically identical.

If an individual stated the change made to the process favorably improved the output based on observation of data sets x_1 and x_2, they would be categorically wrong. Both data sets come from the same distribution of data, and random chance caused the obvious difference in the two averages. This is a slight variation on the bead box effect.

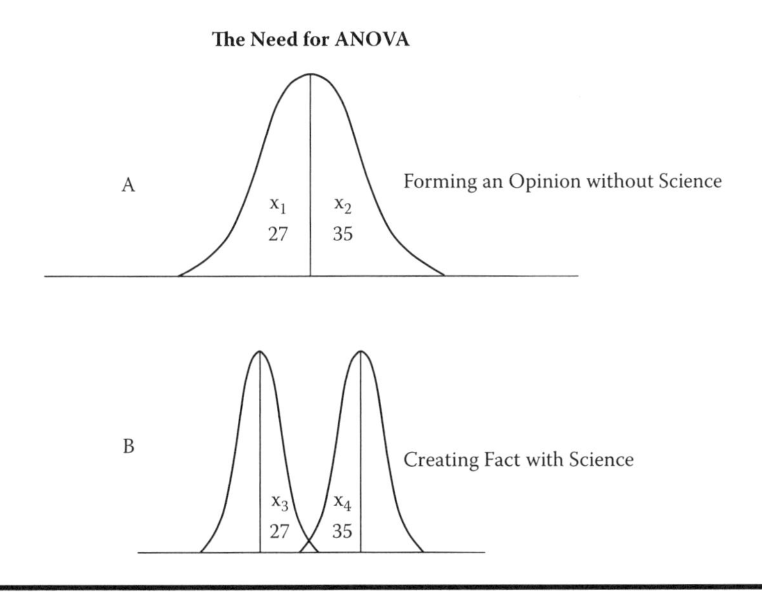

Figure 5.6 The need for analysis of variance (ANOVA).

On the other hand, if an individual were to make the same statement about data sets x_3 and x_4, based on observation of the two data sets, they would be (due to luck) absolutely correct. Data sets x_3 and x_4 come from two different distributions created due to the shift in process average caused by the process change.

Obviously, it is not acceptable to compare two averages without some science in order to draw a conclusion based on fact as opposed to pure judgment.

There is a very simple method provided by Excel that enables anyone to determine if a distribution has shifted and thus represents a process change. The technique called ANOVA has been around for almost 100 years and is rarely used in industry to better understand manufacturing processes. Table 5.5 represents attenuation data from the fiber optics industry. The averages of the two subgroups are 48.48 and 51.36, respectively.

The question is, “Is there a significant difference in the output of the process after a change was made?”

The ability to make an informed decision using the ANOVA screen provided by Excel is very easy:

- Enter the data into an Excel spreadsheet using one column for each data set.
- Under **Tools**, choose **Data Analysis**.
- **ANOVA Single Factor** is already highlighted.
- Click **OK**.
- Highlight the data on the spreadsheet, and it will automatically go to **the Input Range** field.
- Click **OK**.
- A table will appear with important information regarding whether or not a process shift has occurred.

See Table 5.6.

The two values under columns labeled *F* and *F crit* are the only two pieces of information that we need be concerned with.

If the *F* value is less than the *F crit* value, we can be confident that a process average shift has not occurred, and the difference in the two averages is due to random chance—luck of the draw.

In layman’s terms, there is a 95% probability that a process average shift did *not* occur when the process change was made. It is so much more effective to present a case to management in terms of “we are 95% confident a

Table 5.5 Tabled Data for an Analysis of Variance (ANOVA)

	A	*B*
	56.21	55.83
	54.00	43.02
	56.23	60.21
	49.00	59.27
	54.82	51.04
	50.60	53.23
	48.94	48.85
	44.70	54.38
	50.01	42.29
	35.20	53.33
	53.00	52.81
	44.83	53.13
	50.20	52.81
	47.90	51.46
	48.40	48.13
	50.11	52.40
	42.40	47.81
	45.30	50.42
	45.13	50.83
	42.80	48.00
Average	48.48	51.46

process average shift did not take place" as opposed to using phrases such as "we don't *think* a process average shift has occurred."

Of course, in the absence of an ANOVA, with the two averages of 48.48 and 51.46 some people might contend a process average shift had taken place, especially if they were defending their opinion that the process change was necessary.

Decisions regarding instability and other process changes must take into account factual data as opposed to opinion.

Table 5.6 Analysis of Variance (ANOVA) Table

Summary						
Groups	*Count*	*Sum*	*Average*	*Variance*		
Column 1	20	969.78	48.489	27.30098		
Column 2	20	1029.25	51.4625	19.96324		
ANOVA						
Source of Variation	*SS*	*df*	*MS*	*F*	*P-Value*	*F crit*
Between Groups	88.41702	1	88.41702	3.741394	0.060552	4.098169
Within Groups	898.0202	38	23.63211			
Total	986.4372	39				

Notes: If *F* is smaller than *F crit*, there was not a change in the process.
If *F* is larger than *F crit*, there was a change in the process.

Incapability

The lack of capability depends on the position of the process average and the amount of variation about the average. When we discuss problems of incapability, we are usually investigating sources of excessive variation.

Virtually everything written in this chapter regarding how to identify causes of instability, with the exception of ANOVA, will also apply to incapability. Incapability, as mentioned, is an excess of variation, and ANOVA deals only with identifying changes in averages.

When identifying the cause of incapability, it is a good idea to refer to the flow diagram of Figure 3.6 and ensure measurement processes, methods, and condition of equipment have been reviewed. Interviews with operators can address operator technique differences and local environmental concerns. That leaves raw material and equipment parameters once again.

A simple three-factor, two-level experiment can be designed to identify the causes of excessive variation, only this time, the response is the variation of samples chosen, not the average of those samples.

There is a major pitfall encountered by many manufacturing problem solvers when analyzing variation, and it is rooted in using an inappropriate method to determine the degree of variation.

Calculated versus Estimated Standard Deviation

It is common practice to describe the width of a normal curve in terms of standard deviation; some people use the term *sigma*. Outside the world of statisticians, the terms are used interchangeably.

Convention has placed three sigma (or standard deviation) either side of the data average. If a product characteristic is centered at 1 with one sigma of 0.01, we can expect to have product as small as 0.970 (1 – 0.03), and product as large as 1.03 (1 + 0.03). It is a simple concept, and it serves to describe the amount of variation in a standard format.

The formula for calculating standard deviation can be found in the Appendix under "Chapter 5, Calculated Method of Standard Deviation." This formula treats the data from a process that is experiencing process average shifts in the the same way it treats data from a process that is *not* experiencing process average shifts.

Let's assume a process capability study has just been performed. The results are graphically described in Figure 5.7.

Curve A represents the process at the beginning of the study. The product characteristic is centered at 0.05 with one standard deviation of 0.001. With three standard deviations either side of the process average, the process would be yielding product as low as 0.497 and as high as 0.503. During the study, however, the product characteristic average shifted upward to

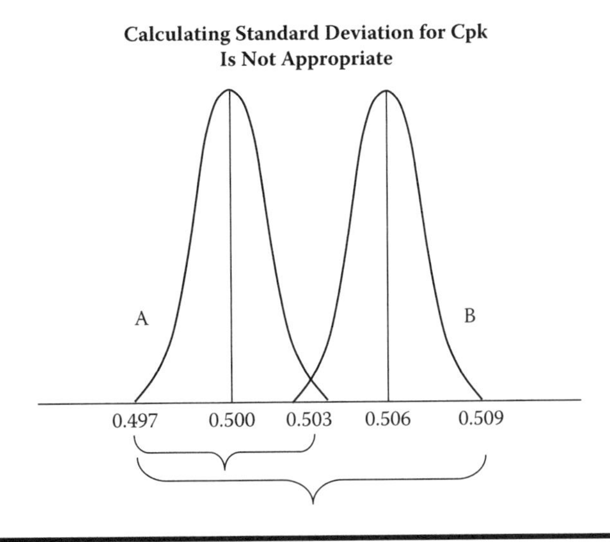

Figure 5.7 Calculating standard deviation for the capability index (Cpk) is not appropriate.

0.506. Please note the amount of variation did not increase when the average increased. Curve B is centered at 0.506 with the same three standard deviations either side of the average.

If we were interested in understanding the total amount of variation of the product being produced, inclusive of the process average shift, it would be appropriate to calculate the standard deviation using the standard formula in the Appendix under "Chapter 5, Calculated Method of Standard Deviation."

However, if we wanted to know the standard deviation of the process for purposes of determining the Cpk, it would not be appropriate to use the calculated method of standard deviation. For Cpk purposes, we must use the method found in the Appendix under "Chapter 5, Estimated Method of Standard Deviation."

For the purpose of process capability studies, we are concerned only with the amount of normal variation, specifically the amount of process variation in the absence of process average shifts. Therefore, the calculated standard deviation *must not* be used in order to determine the Cpk. For purposes of Cpk, the standard deviation must be derived using the estimated method of standard deviation. The arithmetic for this method is found in the Appendix under "Chapter 5, Estimated Method of Standard Deviation."

When calculating standard deviation, Excel uses the total amount of variation, which would include any variation resulting from process average shifts.

Table 5.7 represents data provided for the purpose of demonstrating how different the two methods are. The Excel calculated method of standard deviation yields one standard deviation of 0.0032.

Figure 5.8 represents the Individuals control chart for the same data. Please note the significant process average shift and the one standard deviation value of 0.0012.

The control chart limits, which are plus and minus three sigma about the centerline, are determined using the estimated method of standard deviation value determined as follows, referring to Table 5.7:

- The absolute difference between data point 1 and data point 2 would represent moving range 1: 0.0497 – 0.0487 = 0.001.
- Moving range 2 would be the absolute difference of data points 2 and 3: 0.0487 – 0.0502 = 0.0015. (Because we are taking the absolute value, we ignore the minus sign.)
- Moving range 3 would be the absolute difference of data points 3 and 4: 0.0502 – 0.0513 = 0.0011.

Table 5.7 Data Illustrating Difference between Estimated and Calculated Standard Deviation

0.0497	0.0493	0.0497	0.0517	0.0579	0.0554
0.0487	0.0483	0.0496	0.0484	0.0557	0.0561
0.0502	0.0482	0.0513	0.0556	0.0551	0.0546
0.0513	0.0490	0.0499	0.0559	0.0570	0.0562
0.0512	0.0492	0.0498	0.0570	0.0562	0.0566
0.0517	0.0479	0.0495	0.0554	0.0588	0.0576
0.0478	0.0494	0.0520	0.0569	0.0550	0.0532
0.0498	0.0496	0.0509	0.0557	0.0553	0.0522
0.0511	0.0501	0.0524	0.0555	0.0557	0.0556
0.0489	0.0496	0.0493	0.0554	0.0564	0.0544

Notes: Average range method sigma = average range/1.128.
Excel method sigma = 0.0032 = 0.0014/1.128 = 0.0012.

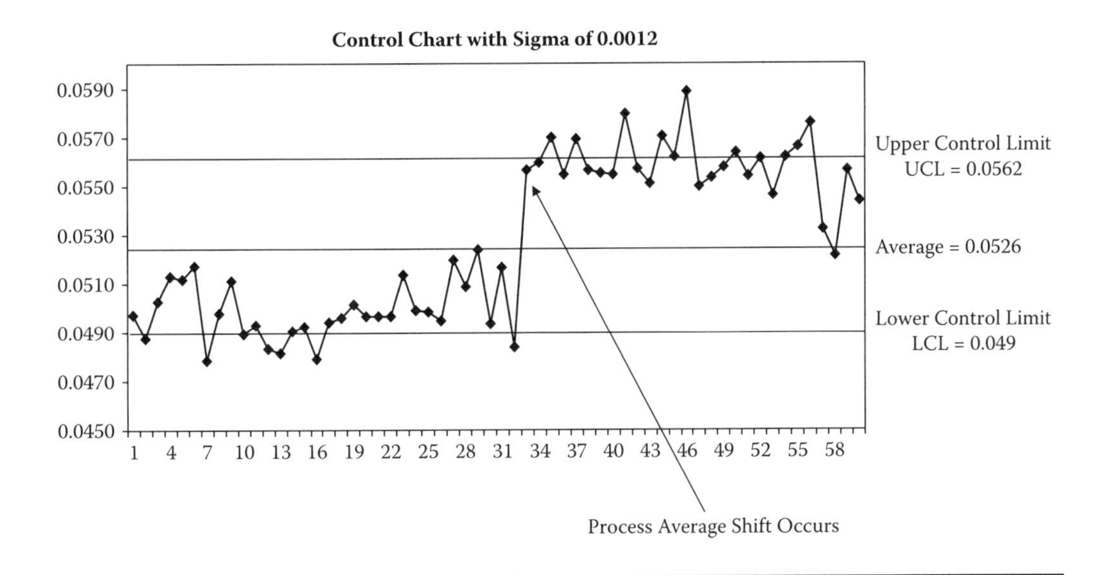

Figure 5.8 Control chart with sigma of 0.0012.

The average of all the moving ranges is divided by a constant value of 1.128 in order to determine one standard deviation of normal variation. The value of 1.128 comes from Table A.1 (Appendix), the constant for a subgroup size of two is used because each moving range is developed using two data points.

In this example, one standard deviation using the estimated method moving range is 0.0012.

The Excel method of calculating standard deviation does not compensate for the process average shift and presents a standard deviation almost three times the standard deviation, which would be needed to determine the Cpk value. The Excel method is totally inappropriate for purposes of process capability studies and calculation of Cpk.

Over the years, I encountered many companies that could not qualify their product for customers such as Ford, GM, and Motorola, because they could not achieve the required Cpk. Their processes were experiencing minor process average shifts due to changes in raw material, voltage fluctuations, and so forth. Because they were using Excel or the wrong statistical software, the minor process average shifts were contributing to a false Cpk value of less than 1.33. As soon as the appropriate method of estimating standard deviation was applied, the Cpk value indicated the processes to be more than capable.

Everyone realizes there is no such thing as a perpetually stable process. All processes experience minor process average shifts. The whole concept of requiring a process to have a Cpk of 1.33 is to allow the process to sustain minor process average shifts while producing product within specification. Therefore, we *must* determine Cpk in the absence of process average shifts. We *must* use the estimated method of standard deviation when analyzing manufacturing processes. This is not a concept that is well understood in industry.

Another Simple Designed Experiment

One last word on identifying root causes of instability and incapability—the three-factor, two-level experiment performed on the missing clip problem was defined as a simple approach to designed experiments, and it required eight runs. There is one tool that is even simpler—the two-factor, two-level experiment that requires only four runs. The matrix for this experiment is found in Table 5.8. This technique would be used when it is desirable to trial only two process parameters using high and low levels.

Table 5.8 A Two-Factor, Two-Level Experimental Matrix

Run	*Factor A*	*Factor B*	*Response*
1	+	+	
2	+	–	
3	–	–	
4	–	+	

Chapter 6

Understanding the Measurement Process

> If you cannot measure it, you cannot improve it.
>
> **Lord Kelvin**

Overview

Particle physicists use the term *observer effect* when they make reference to the fact that measuring a particular phenomenon will change it. One example is that an electron acts like a particle when it is not being observed and assumes the characteristics of a wave function when it is being observed. This idea that we cannot measure something without changing it is often the result of the measurement devices employed which, in some way, alters the physical nature of what is being measured. A more commonplace example of this concept is the result we get when we check the pressure on our automobile tires. The very act of applying the pressure gauge will release some air, thus changing the pressure.

We can easily apply this concept to manufacturing when we consider that when measuring the diameter of compressible rubber feed rolls with a micrometer, the application of the micrometer anvils to the rubber roll would tend to change the diameter. Of course, in manufacturing, there are other facets of the measuring process that contribute to changing the measured item by virtue of measuring it.

As a practical matter, if we consider the five elements of any manufacturing measurement process (see Figure 6.1b) and take into account their combined effect and possible interactions on the measurement output, which would be measurement data, it would be reasonable to conclude that every measurement process has some degree of variation.

Five Elements of the Measurement Process

People

Every measurement process involves people. Different people will prepare samples to be measured using various techniques, which will contribute to measurement variation. As discussed earlier, the same person measuring the same sample several times will often record measurements that are not identical. Two different people might have very different techniques when measuring the same product characteristic, which will also result in measurement variation.

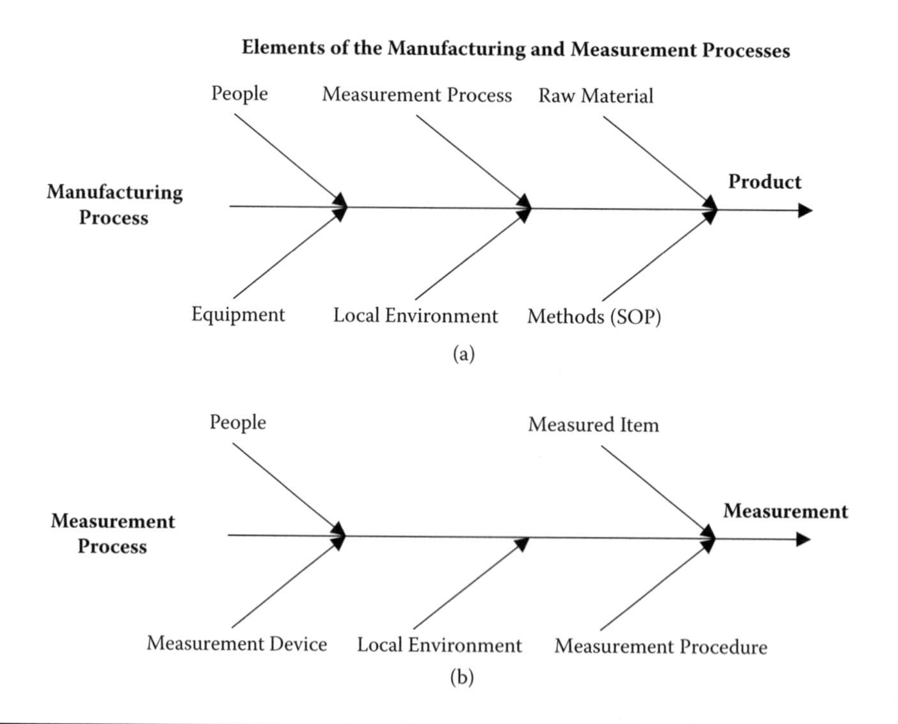

Figure 6.1 Elements of the manufacturing and measurement processes.

Measurement Devices

Regardless of the expense or sophistication of the measurement device, there is always the possibility of interaction between the measurement device and the operator, the measurement device and the product being measured, as well the measurement device and the local environment due to voltage fluctuations, ambient temperature changes, poor lighting, and so forth.

Measured Product

The physical characteristics of the item being measured will often contribute to introducing variation of measurements. Consider the measurement variation resulting from an operator taking several width measurements on a flexible silicone drive belt with a vernier caliper. Multiple measurements taken on a machined steel component with the same vernier caliper would likely evidence less measurement variation.

Measurement Procedure

Measurement procedures are not nearly as common as manufacturing standard operating procedures (SOPs), and when measurement procedures do exist, they rarely offer specific directions for the selection and preparation of samples. Variation in the selection and preparation of samples can be a major contributor to measurement variation.

Local Environment

Poor lighting, vibration, shop floor distractions, and ambient temperature can contribute significantly to measurement variation.

The five elements of the measurement process combine and interact to contribute to the variation of the measurement process output.

Every day, around the world, an unimaginable large number of measurements are performed in manufacturing facilities. Every conceivable product characteristic from the diameters of machined shafts to the tensile strengths of rubber components are measured and often recorded. Numerous process parameters such as pH levels and viscosities are also measured and recorded for the purpose of maintaining the chemistry of various process solutions. It is common for operators and supervisors to select random samples from the output of a process, measure critical characteristics, and (without taking into account the bead box effect) adjust process parameters accordingly. A snapshot of any manufacturing facility would reveal conscientious operators, supervisors, engineers, and managers constantly measuring critical product characteristics with carefully calibrated measurement devices—calibrated devices that may be part of a measurement process that is unstable or incapable with respect to the product characteristic being measured.

A measurement process analysis (MPA) helps us to understand the degree and nature of variation with respect to a defined measurement process. That is, an MPA analyzes the measurement variation resulting from several trained operators (people) using specified measurement equipment (measurement device) and recording multiple measurements taken on the same samples of an identified product characteristic (measured item). The measurements should be made following available written procedures (methods) in the area normally used by the operators for taking measurements (local environment).

The MPA performed on the micrometer in Case Study 3.2, "Crimped Terminal Diameter Customer Returns," disqualified the micrometer with respect to measuring the compressible terminals due to instability of the MPA measurements. Another MPA qualified the noncontact machinist microscope, because the measurements taken with the microscope were stable and capable.

Four Concerns of a Measurement Process Analysis

The recommended MPA protocol addresses three components of the measurement process—precision, operator technique differences, and discrimination. But, as a practical matter, accuracy cannot be ignored when discussing measurement processes.

Accuracy

The dictionary defines *accuracy* as "The ability of a measurement to match the actual value of the quantity being measured." (http://dictionary.com)

In manufacturing, the accuracy of measurement devices is attained by means of calibration programs that are typically part of every manufacturing facility's quality control effort. In many facilities, all the micrometers, vernier calipers, dial indicators, and other handheld measurement devices are periodically collected by a quality department technician and calibrated using standards. For example, gauge blocks certified to the National Bureau of Standards are often used to calibrate micrometers and vernier calipers. If a micrometer reads 0.2505 when measuring a 0.25 gauge block, the technician typically makes adjustments until the micrometer reading agrees with the gauge block designation. After a micrometer is compared to and in agreement with a range of different size gauge blocks, it receives a dated calibration sticker. Gram scales and other major electronic measurement devices are often calibrated by an outside agency, also with the aid of certified standards.

Periodic calibration is necessary. However, it is not in the best interest of a manufacturing facility to rely solely on calibration programs to ensure the efficacy of their measurement processes.

Once a micrometer is calibrated, people are tasked with going onto the shop floor to use the micrometer that was calibrated on a perfectly flat, incompressible, hardened steel block, to make quality and production decisions when measuring manufactured product. These decisions might be made after measuring the lengths of compressible springs, the diameters of flexible electrical connectors, the thickness of compressible rubber drive belts, the diameters of machined shafts that may not be perfectly round, and so forth. There is no reason to believe that a micrometer calibrated on a gauge block will in any way reflect the true dimension of flexible, compressible, imperfectly shaped product. Calibrating a micrometer on a standard ensures accuracy when measuring the standard, but the calibration provides no assurance of the precision of measurements made with the calibrated micrometer on product being produced on the shop floor.

Precision/Repeatability

The dictionary defines *precision* as "The ability of a measurement to be consistently reproduced." (http://dictionary.com)

A micrometer is a simple measurement device until it is picked up by an individual for the purpose of making a measurement; then the micrometer becomes part of a total measurement process. Precision refers to the ability of a measurement device, when operated by a single individual, measuring the same part multiple times in the same location, to record consistent measurements. I used the term *consistent measurements* because it is recognized that it is almost impossible for an individual to repeat any task in exactly the same way multiple times, and this is true when making multiple measurements. When the multiple measurements have a certain level of consistency, we can deem the measurement process as being stable. However, when the measurements exceed that level of required consistency, the measurement process is determined to be unstable and is disqualified from measuring the specific product characteristic in question. The MPA protocol results in a combination of two control charts. One of these control charts—the Range chart—indicates, at a glance, whether or not the measurement process is stable or unstable when measuring the product characteristic in question.

An imprecise (unstable) measurement process used for the purpose of performing a process capability study could very well result in false indications that a stable manufacturing process is unstable. Using an unstable measurement process in production or quality assurance could result in out-of-specification product being shipped and product that meets specification being rejected.

Reproducibility

The dictionary defines *reproducibility* as "To produce again or anew; re-create." (http://dictionary.com)

It would be most desirable for a measurement process to reproduce the same results regardless of the person performing the measurement; unfortunately, this is not always the case. Anyone who has been in manufacturing for any length of time has undoubtedly observed two individuals debating the actual measurement of a characteristic that has been taken on a sample of product—one insisting the sample was just out of specification (an inspector) and the other insisting the sample was just within specification (a production supervisor).

The effect operator technique differences have on the outcome of the measurement process is termed *reproducibility*. An MPA gauges the reproducibility of a measurement process by comparing the averages of multiple measurements recorded by different operators when measuring the same samples using the same measurement device. If the averages of the operators' multiple measurements are significantly different, there is a lack of reproducibility affecting the measurements.

Discrimination

The dictionary defines *discrimination* as "The ability or power to make fine distinctions." (http://dictionary.com)

As mentioned in Chapter 3, if the measurement variation is equal to or greater than the amount of normal variation of the product being measured, the measurement process will not be able to tell the difference between some samples, and the ability to tell the difference between samples is critical to the study of process variation.

Variable process capability studies always generate control charts (X Bar and R charts and Individual's charts are the most common), and the control chart limits are a function of the average range between measured samples. If the measurement process cannot adequately discriminate between samples, the number of range opportunities will be limited, which will result in an artificially low average range causing the control chart limits to be more narrow than they should be. Narrow control chart limits can result in false indications of manufacturing process instability.

Statistical product control charts, the analysis of variance (ANOVA) technique, and designed experiments, important tools in the identification of the root causes of instability, are based on the ability to identify the normal variation of the measured product. If a measurement process does not have adequate discrimination, it will not have the capability to tell the difference between some samples made during periods of normal variation. As a result, statistical product control chart data patterns with respect to the control chart limits, and results from ANOVAs and designed experiments will be misleading. If we have no reliable estimate of normal variation of product due to a lack of discrimination, we cannot successfully employ the tools mentioned above to understand and identify causes of instability.

Due to the requirements of precision, reproducibility, and discrimination, a measurement process must be qualified on the product characteristic it is intended to measure.

CASE STUDY 6.1 WIRE DIAMETER MICROMETER AND MICROSCOPE MEASUREMENT PROCESS ANALYSES (MPAs)

A nondestruct measurement process involves the measurement of a sample that does not destroy or change the characteristic of the sample being measured. Measuring the length of a mattress spring assembly with a tape measure or the diameter of a shaft with a micrometer are both examples of nondestruct measurement processes.

A manufacturer of high-tech wire for the aerospace industry was receiving complaints and returns from a very large and valued customer due to out-of-diameter specification on a specific single end insulated wire.

The first step in the problem-solving exercise is to determine if the problem is one of instability or incapability by means of performing process capability studies. Of course, if the flow diagram of Figure 3.6 is to be followed, the second step in performing a process capability is to perform an MPA using the type of micrometer that would be used to measure the product. Both production and the quality group had been using a Brown & Sharp thimble micrometer with resolution of 0.0001.

MPA SAMPLE SELECTION

Technicians were instructed to collect five 6-inch pieces of wire from five different spools produced over an 8-hour shift. Probably one of the most important tasks in performing an MPA is the collection of the samples. The ability to discriminate between samples made during periods of normal variation is a necessary element of any measurement process that is to be used for purposes of process analysis. In order to determine if a measurement process can distinguish between samples made during periods of normal variation, it is necessary to choose the samples over a period of time that is believed will represent normal variation. If the five 6-inch-long samples were selected from the end of one spool of finished product, it is very doubtful the samples would represent normal variation, and the measurement process would be erroneously disqualified from measuring the diameter of this product due to a perceived lack of ability to discriminate. Engineers, operators, and supervisors who are familiar with the manufacturing process are generally very capable in defining over what period of time the samples should be selected.

The location to be measured on each sample was marked with indelible ink, and each of the samples was tagged from one to five.

One engineer accepted responsibility to be sample coordinator. She was instructed to give the five samples in turn to each one of the four operators participating in the study. Operator 1 would receive the samples and the designated micrometer at his workstation where he normally measured product. Operator 1 was instructed to measure the sample at the designated location, record each sample measurement on a piece of paper, and return the samples, the micrometer, and the recorded measurements to the coordinator. The coordinator would then pass the samples and the micrometer onto operator 2 at her workstation with the same instructions, then operator 3 and then operator 4.

Toward the end of the shift, the coordinating engineer interchanged the sample tags, made a record of which newly tagged samples were the original samples one through five, and repeated the round-robin measurement exercise among the four operators, explaining to each operator the second round of samples were different than the first round. The coordinator was instructed to interchange the samples, because it is prudent to ensure the operators are not biased and attempt to “force” samples one through five to measure, on the second round, what they measured earlier in the day. Measurements are very personal, and some operators will believe it is a shortcoming on their part if they cannot exactly repeat measurements when measuring the same item more than once.

The next day, early in the shift, the coordinator repeated the exercise, making certain the tags were interchanged once again. Then there was another round-robin at the end of the shift and one more the next day for a total of five sets of measurements from each of the four operators.

The total amount of time for the study to be performed was on the order of 20 minutes. The five round-robins were performed at each operator’s workstation, stretched out over 3 days and performed at different times of the day to ensure the normal shop floor environment was in play during the study. Some organizations have been known to perform an MPA by engaging engineers in the task of measuring samples in an air-conditioned, well-lit training room instead of asking operators to participate on the shop floor where measurements are usually performed. Involving nonoperators in measuring samples under artificial conditions might provide better results but not necessarily the results that reflect shop floor reality.

When the data were compiled, four operators had measured five samples five times, each over 3 days. The data were entered into the Excel spreadsheet on Table 6.1.

Note the data consist of 20 subgroups with five measurements in each subgroup. Also note the average and range of each subgroup appear at the bottom of each column. These averages and ranges are used to create the X Bar (Average) chart and Range (R) chart found in Figure 6.2.

RANGE CHART

The Range chart serves only one purpose when analyzing a measurement process. Specifically, the Range chart is the indicator of stability. Each data point on the Range chart represents the range of one subgroup from Table 6.1 in sequential order. And each of the range values is an indication of the precision or repeatability of the measurement process (micrometer, operator, measured item, method of measurement, and local environment). This concept of measurement process is being reinforced due to the very human tendency to attribute a lack of precision solely to the operator.

If, on the Range chart, all the values fall within the upper control limit, we can categorically state the measurement process is stable. Do the range values indicate different measurements recorded by the same operator on the same sample? Yes. However, the differences (ranges), because they all fall within the upper control limit, demonstrate a certain consistency; therefore, this measurement process is stable.

If one Range value were to be above the Range chart upper control limit, it would be an indication that the measurement process is unstable, and the cause of the instability would have to be identified and removed, and the study would have to be performed again.

X BAR CHART

The X Bar chart serves two purposes when it is being applied to the analysis of a measurement process.

Operator Technique Differences

The patterns of the averages on the X Bar chart are a visual representation of the measurement process reproducibility. In a perfect world, the pattern

Table 6.1 Measurement System Analysis for Nondestructive Test

Part Number and Name	Insulated Wire	Device Name	Micrometer	Date	January 2006
Characteristic	Diameter	Device Number	QA 102	Performed By	Andy P.

Sample Size (n)	5	*n ranges from 1 to 5*
Number of Repetitions (r)	5	*r ranges from 1 to 5*
Number of Operators (o)	4	*o ranges from 2 to 4*

	Operator 1					*Operator 2*				
	Sample 1	*Sample 2*	*Sample 3*	*Sample 4*	*Sample 5*	*Sample 1*	*Sample 2*	*Sample 3*	*Sample 4*	*Sample 5*
Repetition 1	0.0658	0.0653	0.0657	0.0653	0.0656	0.0635	0.0625	0.061	0.0611	0.0618
Repetition 2	0.065	0.0647	0.0656	0.0649	0.0649	0.0628	0.0622	0.0618	0.0624	0.061
Repetition 3	0.638	0.0643	0.0656	0.0651	0.0652	0.0621	0.0612	0.0628	0.0617	0.062
Repetition 4	0.0642	0.0634	0.0634	0.0635	0.0637	0.0618	0.0618	0.0622	0.0628	0.0626
Repetition 5	0.0625	0.0625	0.0651	0.0656	0.0656	0.0612	0.0618	0.0618	0.0622	0.0628

Average	0.064	0.064	0.065	0.065	0.065	0.062	0.062	0.062	0.062	0.062
Range	0.0033	0.0028	0.0023	0.0021	0.0019	0.0023	0.0013	0.0018	0.0017	0.0018

	Operator 3					Operator 4				
	Sample 1	*Sample 2*	*Sample 3*	*Sample 4*	*Sample 5*	*Sample 1*	*Sample 2*	*Sample 3*	*Sample 4*	*Sample 5*
Repetition 1	0.0625	0.0635	0.0623	0.0614	0.0614	0.0639	0.0645	0.0621	0.0639	0.0639
Repetition 2	0.0605	0.062	0.0632	0.0626	0.063	0.0628	0.0642	0.0638	0.0633	0.0633
Repetition 3	0.0625	0.063	0.06202	0.0629	0.0596	0.062	0.0634	0.0633	0.0634	0.0628
Repetition 4	0.0644	0.0605	0.062	0.0626	0.0626	0.061	0.0616	0.0616	0.0613	0.0617
Repetition 5	0.061	0.064	0.0604	0.0643	0.0613	0.062	0.0626	0.0625	0.0617	0.0621

Average	0.06218	0.0626	0.06198	0.06276	0.06158	0.06234	0.06326	0.06266	0.06272	0.06276
Range	0.0039	0.0035	0.0028	0.0029	0.0034	0.0029	0.0029	0.0022	0.0026	0.0022

formed by the first five averages from operator 1 would be identical to the patterns formed by the other three operators' averages, indicating there was little to no operator technique differences. In this instance, there appear to be significant differences between the patterns. The X Bar chart indicates operator 1 is measuring the product approximately 0.003 higher than the other three operators. This does not suggest operator 1's measurements are wrong—they are just different. Also, the pattern of the averages for operator 2 shows much less variation between those averages than we see with the other operators. This does not mean that the averages of operator 2 are wrong, they are just different than those of the other three operators.

Precision

When the X Bar chart control limits are so wide that they encompass a majority of the averages, it is a visual indication the measurement process is being dominated by error due to a lack of precision or repeatability. The two rows in Table 6.1 labeled Range provide insight into the amount of imprecision that exists when this product diameter is being measured with the designated handheld micrometer. The ranges in each of the 20 columns represent the differences recorded by the operators when measuring the same sample in the same location with the same micrometer five times over a period of 3 days. Note that operator 3, sample 1, registered the highest range of 0.0039. The good news is that the measurement process is stable due to all the ranges being within the Range chart upper control limit; the bad news is that the consistent ranges have excessive variation causing the measurements to be dominated by error. This measurement process is stable (based on the Range chart patterns) but not capable (based on the X Bar chart patterns and wide control chart limits).

The bar graphs and the discrimination ratio in Table 6.2 provide numerical translations of the X Bar chart.

Of the total amount of variation generated by the measurement process, the bar graphs indicate

44.89% is due to the imprecision or lack of repeatability.
54.55% is due to differences between the operators or lack of reproducibility.
0.56% is due to the product variation.

Please note the discrimination ratio above the bar graphs is 1. This is a numerical translation of the X Bar chart control limits, which, as mentioned, are very wide. A discrimination ratio of 1, in broad terms, indicates the variation of the measurements is about equal to or greater than the variation of the product. This measurement process cannot distinguish between many of the samples.

A meeting was conducted which included engineers, supervisors, production, quality management, and the operators who participated in the measurement study. Table 6.1, Figure 6.2, and Table 6.2 were presented to the group.

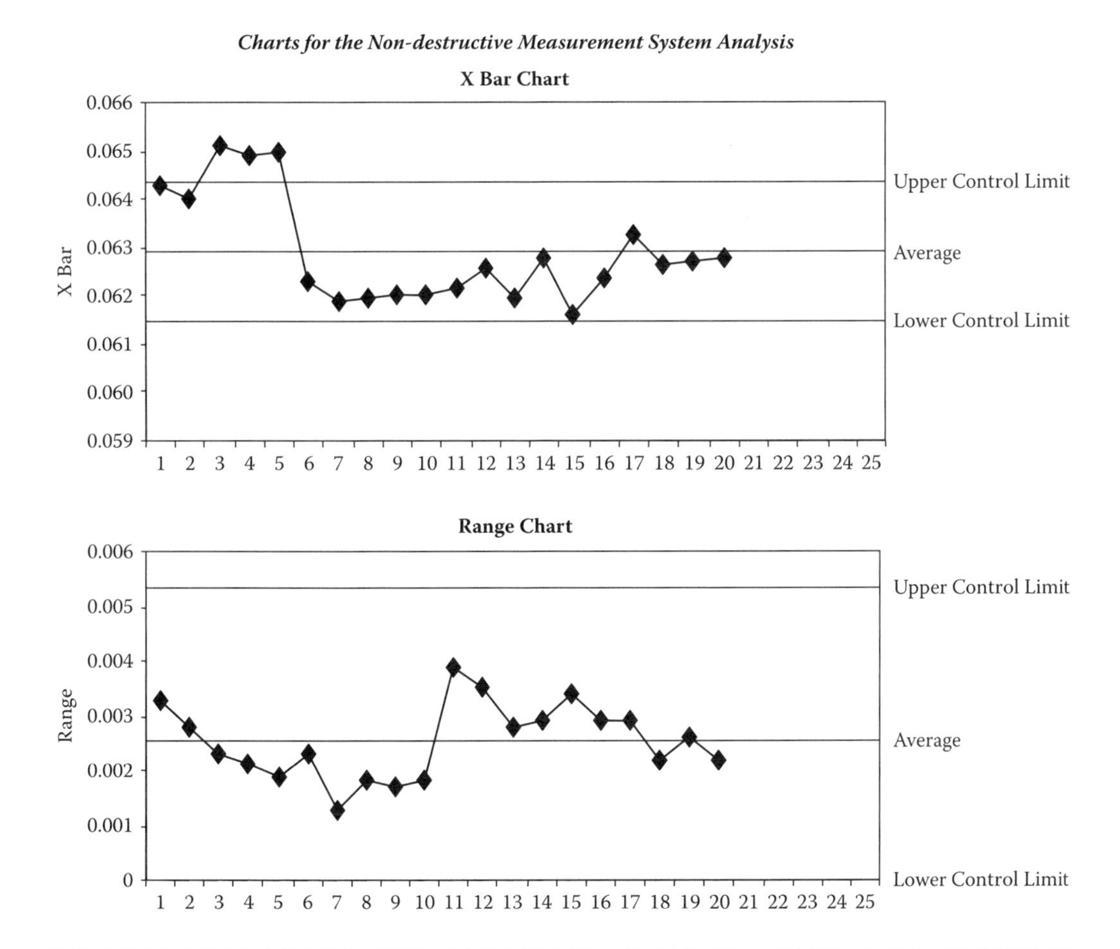

Figure 6.2 Charts for the nondestructive measurement system analysis.

Table 6.2 Components of Measurement System Variation

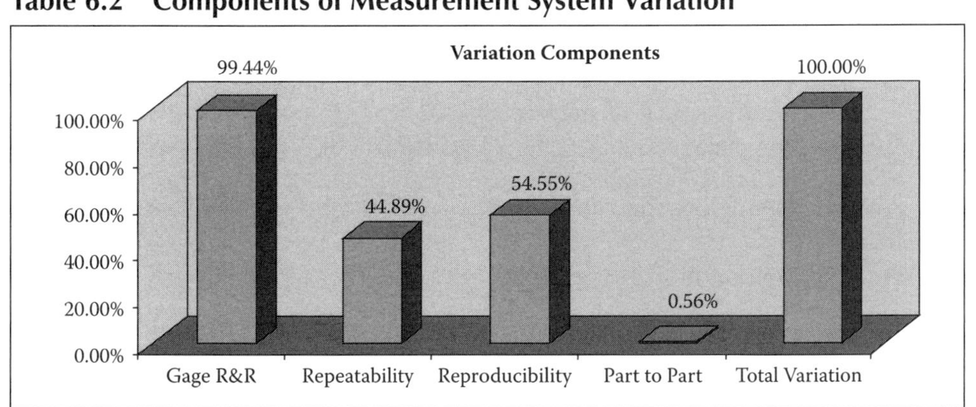

Before reviewing the data, it was pointed out that the operators' names were not recorded in Table 6.1 and that operator 1, for example, was not the first operator who measured the samples. It was also emphasized that the operator was only one of five elements that contribute to the measurement variation resulting from every measurement process.

When Table 6.1 was projected on a screen, the operators expressed surprise at the degree of imprecision resulting from the same person measuring the same product with the same micrometer in the same location five times. These comments are not uncommon when reviewing MPA results.

Reviewing the charts and bar graphs, everyone agreed that the measurement process was stable, and there were obvious operator technique differences. It was also agreed that the diameter measurements that had been relied on for years to make production and quality decisions for this product were totally dominated by measurement process error. Everyone also agreed that most of the measurement variation was the result of the contact nature of the micrometer on the compressible insulation. A non-contact method of measuring the diameter of the extruded product was necessary.

A machinist microscope (jeweler's microscope) was suggested—reminiscent of Case Study 3.2, "Crimped Terminal Diameter Customer Returns."

A machinist microscope is like any other microscope except for the fact that the small table on which the sample is placed can be moved in the X-axis and Y-axis by means of two micrometer barrel dials. Also, a fine crosshair has been etched onto the left eyepiece.

A wire sample can be placed onto the sample table, and one side of the wire can be aligned with the crosshair. The relative position of the table is recorded from the appropriate micrometer dial reading, the micrometer barrel is then rotated until the other side of the wire is aligned with the crosshair, and a second reading from the micrometer is recorded. The difference between the two micrometer readings is, in this case, the diameter of the sample.

Caution: Machinists' microscopes can be imprecise due to poor lighting and gear slop on the micrometer dials. A 360° circumferential light source is recommended. Also, if the operator overshoots the crosshair backing up to "the target" may introduce variation due to gear slop. Always close in on the target.

A machinist microscope was acquired on loan from a retail company specializing in inspection equipment, and a second MPA was organized using the same samples and the same operators.

The operators were engaged in four more round-robin measurement studies using the same samples. Table 6.3, Figure 6.3, and Table 6.4 illustrate the data and the resulting analysis. For expediency sake, only four repetitions were performed on the second study.

Table 6.3 Measurement System Analysis for Nondestructive Test

Part Number and Name	Insulated Wire	Device Name	Microscope	Date	January 2006
Characteristic	Diameter	Device Number	On Loan	Performed By	Andy P.

Sample Size (n)	5	*n ranges from 1 to 5*
Number of Repetitions (r)	4	*r ranges from 1 to 5*
Number of Operators (o)	4	*o ranges from 2 to 4*

	Operator 1					*Operator 2*				
	Sample 1	*Sample 2*	*Sample 3*	*Sample 4*	*Sample 5*	*Sample 1*	*Sample 2*	*Sample 3*	*Sample 4*	*Sample 5*
Repetition 1	0.0655	0.0626	0.0625	0.0622	0.0641	0.0654	0.0634	0.0635	0.0622	0.0641
Repetition 2	0.0649	0.0632	0.0621	0.0619	0.0642	0.0657	0.0629	0.0641	0.0628	0.0651
Repetition 3	0.0647	0.0635	0.0619	0.0618	0.064	0.0658	0.0631	0.0637	0.0627	0.0653
Repetition 4	0.0653	0.0629	0.0623	0.0617	0.0635	0.066	0.0639	0.0639	0.0626	0.0647
Repetition 5										

Average	0.065	0.063	0.062	0.062	0.064	0.066	0.063	0.064	0.063	0.065
Range	0.0008	0.0009	0.0006	0.0005	0.0007	0.0006	0.001	0.0006	0.0006	0.0012

(Continued)

Table 6.3 Measurement System Analysis for Nondestructive Test (Continued)

	Operator 3					*Operator 4*				
	Sample 1	*Sample 2*	*Sample 3*	*Sample 4*	*Sample 5*	*Sample 1*	*Sample 2*	*Sample 3*	*Sample 4*	*Sample 5*
Repetition 1	0.0661	0.0637	0.0638	0.0629	0.0657	0.0669	0.0626	0.062	0.0633	0.0642
Repetition 2	0.0666	0.0627	0.0632	0.0628	0.0656	0.0662	0.0626	0.0634	0.0623	0.0648
Repetition 3	0.0664	0.0637	0.064	0.062	0.0645	0.0663	0.0634	0.0629	0.0627	0.0642
Repetition 4	0.0667	0.0642	0.0642	0.0625	0.0647	0.0664	0.0631	0.0625	0.063	0.0643
Repetition 5										
Average	0.06645	0.06358	0.0638	0.06255	0.06513	0.06645	0.06293	0.0627	0.06283	0.06438
Range	0.0006	0.0015	0.001	0.0009	0.0012	0.0007	0.0008	0.0014	0.001	0.0006

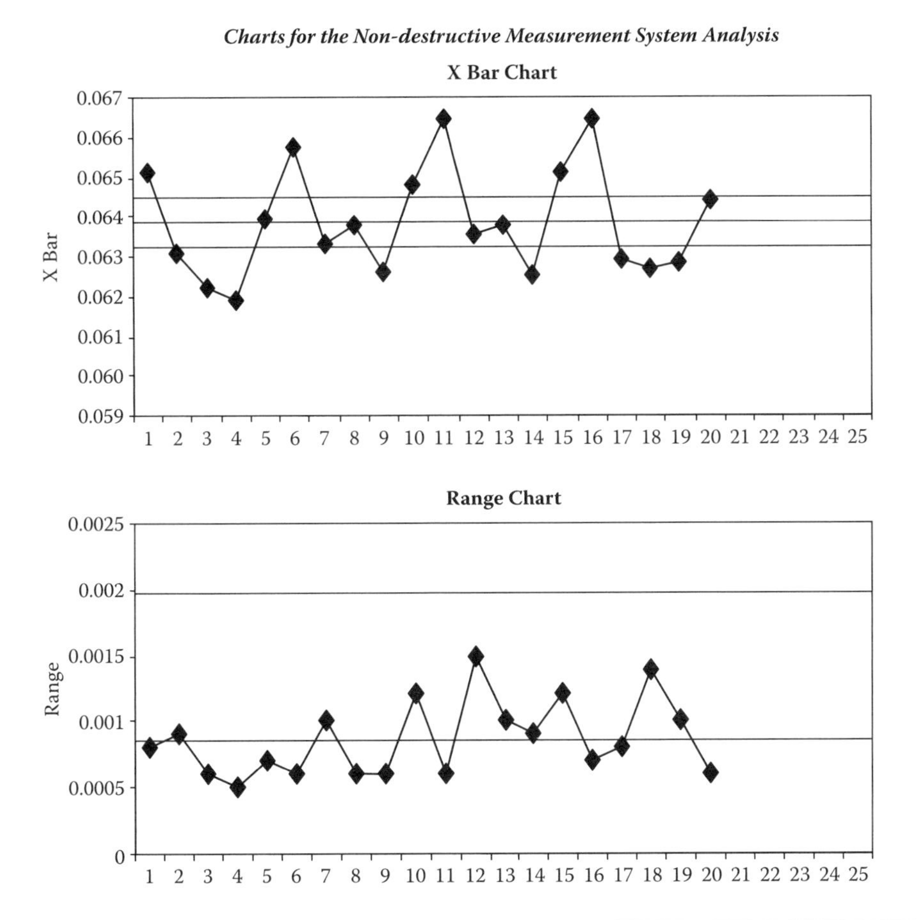

Figure 6.3 Charts for the nondestructive measurement system analysis.

Table 6.4 Components of Measurement System Variation

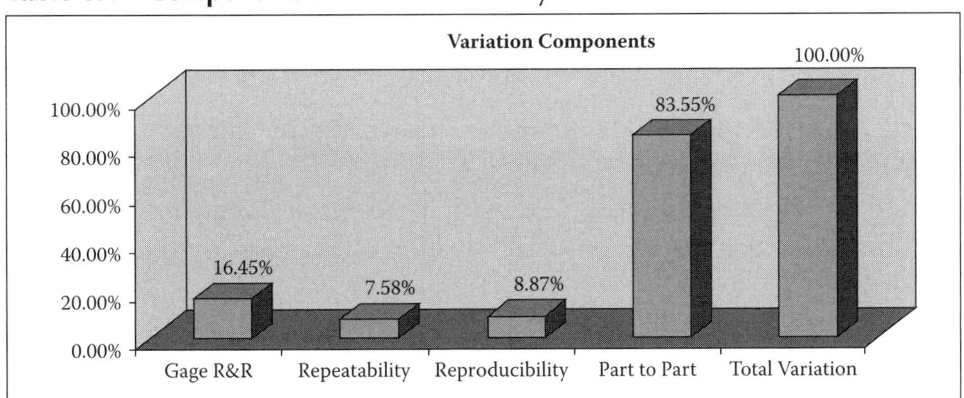

After the study was complete, a review session with the involved parties was once again conducted. Even before Table 6.3 was projected on a screen, the operators announced the measurements for this study had been "way more consistent."

The chart patterns on Figure 6.3 look very different than the micrometer chart patterns of Figure 6.2 from the first study. The Range chart informs us the microscope measurement process is stable, as it was for the micrometer. The narrowness of the control limits of the X Bar chart indicates the measurements are not being dominated by error as they were with the micrometer. Also, the patterns of the averages on the X Bar chart indicating operator technique differences are more similar than they were with the micrometer measurements.

The bar graphs and the discrimination ratio in Table 6.2 put numbers to the visual conclusions derived from the X Bar chart.

Using the microscope, the repeatability and reproducibility are 7.5% and 8.87%, respectively. The part-to-part variation is now 83.55%, which provides a discrimination ratio of 5.

In order to use a measurement process for process analysis, a discrimination ratio equal to or greater than 4 is required.

This microscope measurement process could be used for the planned process capability study for this product, because the microscope measurement process is stable and capable with respect to this product.

Note: The microscope was purchased by the manufacturer. A 360° circumferential fluorescent light source was added to the barrel of the microscope, and another MPA was conducted. With the addition of the light source and standardizing operator technique of closing on the target, minor operator technique differences were further reduced, and the discrimination ratio increased to 6.

Destruct Measurement Process

A destruct measurement study results in destroying the sample being tested. Measuring elongation on an elongation tester is an example of a destruct measurement, because in order to test the percent of elongation, it is necessary to break the sample under load. Strip force testing is another example of a destruct test—for purposes of performing an MPA, a sample can only be stripped from a substrate once.

The performance and analysis of a destruct measurement process is virtually identical to the analysis of a nondestruct measurement process with the exception of sample selection and control chart analysis.

Destruct Measurement Process Analysis (MPA) Sample Selection

When we test a 6-inch-long sample of extruded polymeric material for tensile or elongation, by breaking it on an Instron or some other measuring device, we cannot test the same sample again.

We can, however, make the assumption that a 48-inch-long piece of polymeric material, extruded in less than several seconds, has a minimal amount of elongation variation when compared to the amount of variation realized over an entire production run.

This is a reasonably safe assumption given what we know about the elements of the manufacturing process and their effects on normal variation. Absent some very unusual and severe form of instability in the manufacturing process, it is not probable the operator, the raw material, the equipment, or the local environment would introduce a great deal of variation into the product during the few seconds it takes to produce 48 inches of extruded material. This 48 inches of product would be designated as sample number one, cut into the eight 6-inch segments that would be required for four operators to test and retest the "same sample number one." Each of the "same samples" will be tested only twice to minimize the test–retest error that might result from measuring the "same sample five times," which would be normal practice when performing a nondestruct measurement study.

Of course, to ensure that we provide the measurement process the opportunity to measure a range of five samples that represent normal product variation, the 48-inch-long sample number two would be selected several hours after the 48-inch-long sample number one is chosen. And several hours after sample two is selected, a 48-inch-long sample three would be selected, and so on.

Destruct MPA Control Chart Analysis

We can help to minimize the amount of product variation within a "same sample" by containing each of the five samples to a very finite length such as 48 inches. We can also help to minimize the effect of product variation within a "same sample" by limiting the number of "same sample" measurements to two. However, we must assume that the difference between two measurements of any of the five "same samples" will be inflated to some degree due to product variation and, in some cases, by the preparation of the "same samples."

Preparation of the "Same Samples"

In order to prepare polymeric samples for tensile or elongation testing, it is often necessary to strip the two "same samples" from a substrate; in so doing, the tensile strength of the samples and other physical characteristics can be significantly altered. It is probable the physical characteristics of the two samples will not be altered in exactly the same way, which will increase the difference in the test–retest measurements.

Because of the potential for artificially inflating test–retest error when performing a destruct MPA due to the material variation between two "same samples" and sample preparation, it is recommended the Individuals and Range control charts be used for analysis instead of the X Bar and Range control charts that are employed when performing a nondestruct measurement study. This recommendation stems from the fact that the upper and lower control limits of the Individuals control chart are based on the range between samples (MR) as opposed to the upper and lower control limits of the X Bar chart which are based on the average range (R) of the test–retest data.

Therefore, the Individuals control chart when applied to a destruct measurement process is not a visual representation of the measurement error as is the X Bar chart when applied to a nondestruct measurement error. The Individuals control chart when employed in a destruct MPA is used solely to ensure the manufacturing process was stable during the time the samples were selected.

The Range chart for the destruct MPA is used in the same manner as the Range chart for the nondestruct MPA. Specifically, the Range chart when applied to either a nondestruct or a destruct MPA is used to determine the stability of the measurement process.

To summarize, when analyzing a destruct measurement process, we construct an Individuals chart to ensure the manufacturing process was stable during the sample selection, and we use a Range chart to determine if the measurement process is stable.

CASE STUDY 6.2 ELONGATION TESTER QUALIFICATION

A major manufacturer of polymeric-coated product required shop floor personnel to periodically test percent elongation of the polymeric material at initial process setup and several times during each shift. Elongation measurements were carefully recorded at the workstations and periodically collected and filed with quality control records. Between the shop floor, quality control, the materials laboratory, and research and development (R&D), approximately 80 elongation testers were in use.

A supplier of specialized test equipment contacted the manufacturer and announced the introduction of a new elongation tester with improved resolution.

Before the decision was made to expend a significant amount of money to replace the existing elongation testers with the newer model, it was strongly recommended that one new model be loaned to the manufacturer for a period of several days. During this period of time, an MPA would be performed on the new model and compared to an MPA performed on one of the existing elongation testers.

Five 96-inch-long pieces of extruded material were selected from one extruder over a period of an 8-hour shift. Each 96-inch sample was cut in half and separated into those samples to be tested on the new model and those samples to be tested on the existing model.

From this point on, the procedure for a Destruct MPA is identical to a Nondestruct MPA. Samples are tagged, provided to operators, tested, samples are retagged, retested, and so on.

Of course, the coordinator keeps all the data properly aligned and, in the case of a Destruct MPA, each sample is tested only twice for the reasons previously stated.

Table 6.5 contains the data for both the existing tester model and the new model on loan.

Figure 6.4 illustrates the charts for the two studies. As a reminder—when performing a destruct MPA, the procedure utilizes a combination of the Individuals and Range charts instead of the X Bar and Range chart combination used for nondestruct MPA studies.

The Range chart is used to determine stability in the same way it is used in the nondestruct MPA procedure. For instance, if a range value is above the upper control limit, we have an unstable measurement process. In this case study, both measurement processes are stable.

The Individuals Charts indicate stability of the manufacturing process during the time samples selected and the Range charts indicate both measurement processes are stable.

The Repeatability or Precision of the two measurement processes can be found in Table 6.6

- The existing elongation tester demonstrated a standard deviation of measurement equal to 18.2181.
- The on loan elongation tester demonstrated a standard deviation of measurement equal to 17.0213.

There was no statistically significant difference between the existing tester and the new model therefore management made the decision not to

Table 6.5 Measurement System Analysis for Destructive Test

Part Number and Name	CS2203	Device Name	QA Tester	Date	06 03
Characteristic	Elongation	Device Number	156B	Performed By	George H.

Sample Size (n)	5	*n ranges from 1 to 5*	
Number of Repetitions (r)	2	*r equals 2*	
Number of Operators (o)	4	*o ranges from 2 to 4*	

Existing Tester										
	Operator 1					*Operator 2*				
	Sample 1	*Sample 2*	*Sample 3*	*Sample 4*	*Sample 5*	*Sample 1*	*Sample 2*	*Sample 3*	*Sample 4*	*Sample 5*
Repetition 1	194	155	164	192	144	166	160	171	157	188
Repetition 2	189	147	202	178	163	156	183	160	200	178

Average	191.500	151.000	183.000	185.000	153.500	161.000	171.500	165.500	178.500	183.000
Range	5	8	38	14	19	10	23	11	43	10
Moving Range	40.500	32.000	2.000	31.500	7.500	10.500	6.000	13.000	4.500	9.000

	Operator 3					*Operator 4*				
	Sample 1	*Sample 2*	*Sample 3*	*Sample 4*	*Sample 5*	*Sample 1*	*Sample 2*	*Sample 3*	*Sample 4*	*Sample 5*
Repetition 1	168	168	171	179	180	154	187	174	181	186
Repetition 2	180	203	211	161	193	161	171	202	152	154

Average	174.000	185.500	191.000	170.000	186.500	157.500	179.000	188.000	166.500	170.000
Range	12	35	40	18	13	7	16	28	29	32
Moving Range	11.500	5.500	21.000	16.500	29.000	21.500	9.000	21.500	3.500	

Measurement System Analysis for Destructive Test					
Part Number and Name	CO2203	Device Name	Loaner	Date	06 03
Characteristic	Elongation	Device Number	NA	Performed By	George H.

Sample Size (n)	5	*n ranges from 1 to 5*
Number of Repetitions (r)	2	*r equals 2*
Number of Operators (o)	4	*o ranges from 2 to 4*

(Continued)

Table 6.5 Measurement System Analysis for Destructive Test (Continued)

On Loan Tester	Operator 1					Operator 2				
	Sample 1	*Sample 2*	*Sample 3*	*Sample 4*	*Sample 5*	*Sample 1*	*Sample 2*	*Sample 3*	*Sample 4*	*Sample 5*
Repetition 1	174	165	164	193	166	150	140	152	175	162
Repetition 2	193	147	194	207	190	158	155	156	200	172

Average	183.500	156.000	179.000	200.000	178.000	154.000	147.500	154.000	187.500	167.000
Range	19	18	30	14	24	8	15	4	25	10
Moving Range	27.500	23.000	21.000	22.000	24.000	6.500	6.500	33.500	20.500	1.500

	Operator 3					Operator 4				
	Sample 1	*Sample 2*	*Sample 3*	*Sample 4*	*Sample 5*	*Sample 1*	*Sample 2*	*Sample 3*	*Sample 4*	*Sample 5*
Repetition 1	162	163	168	170	186	170	180	190	180	160
Repetition 2	175	194	172	202	213	205	195	155	165	170

Average	168.500	178.500	170.000	186.000	199.500	187.500	187.500	172.500	172.500	165.000
Range	13	31	4	32	27	35	15	35	15	10
Moving Range	10.000	8.500	16.000	13.500	12.000	0.000	15.000	0.000	7.500	

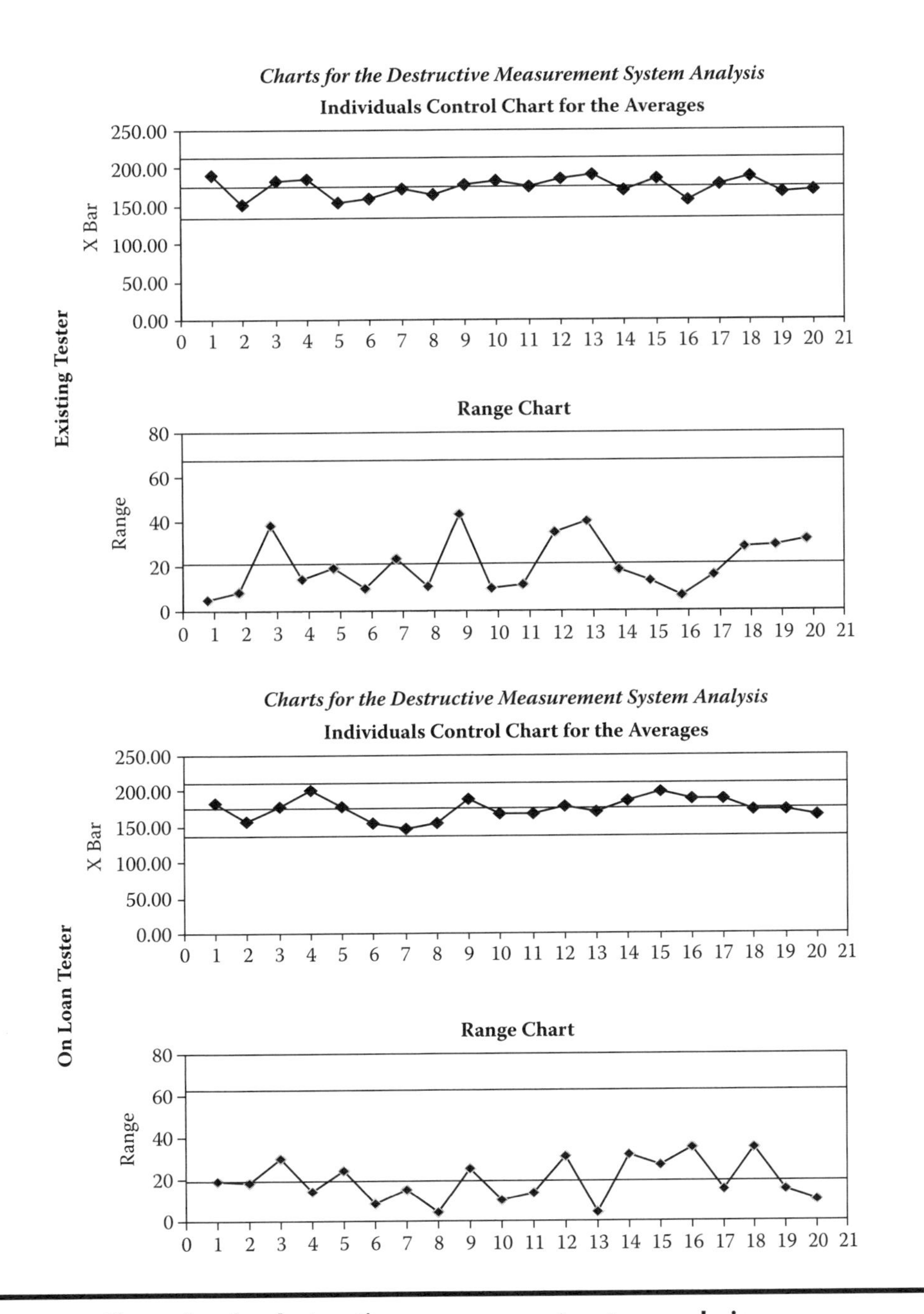

Figure 6.4 Charts for the destructive measurement system analysis.

purchase the new equipment and realized a cost avoidance of more than $150,000.

The mathematics that supports the analysis of a measurement process is far more complex than the simple arithmetic necessary to gain a full understanding of the process capability study technology. In this chapter, I endeavored to provide the reader with the ability to qualify measurement

Table 6.6 Lucent Variation Component Table

Existing Tester			
Variation components	*Std. Dev.*	*Variance*	
Variation of the instrument	18.2181	331.8986	
Variation of the product	4.3585	18.9969	
Variation of the observations	17.5900	309.4082	
Variation of the operators	4.5654	20.8425	

	Std. Dev.	*Variance*	*%Contribution*
Gauge R&R	18.7814	352.7411	94.89%
Repeatability	18.2181	331.8986	89.28%
Reproducibility	4.5654	20.8425	5.61%
Part to part	4.3585	18.9969	5.11%
Total variation	19.2805	371.7380	100.00%

On Loan Tester			
Variation components	*Std. Dev.*	*Variance*	
Variation of the instrument	17.0213	289.7239	
Variation of the product	7.6769	58.9344	
Variation of the observations	17.6761	312.4427	
Variation of the operators	8.5903	73.7933	

	Std. Dev.	*Variance*	*%Contribution*
Gauge R&R	19.0661	363.5172	86.05%
Repeatability	17.0213	289.7239	68.58%
Reproducibility	8.5903	73.7933	17.47%
Part to part	7.6769	58.9344	13.95%
Total variation	20.5536	422.4516	100.00%

processes prior to performing subsequent studies. To make this process as easy as possible, the reader can log onto www.dougrelyea.com and request an electronic copy of the Excel spreadsheets illustrated in Figure 6.3 (nondestruct) and Figure 6.4 (destruct). Once the user has filled in his or her data, with one click the control charts, bar graphs, and related data will be generated and appear as represented in this chapter.

And at risk of being repetitive, I will once again affirm that conclusions drawn from any process analysis performed without a qualified measurement process are immediately suspect.

Chapter 7

Measurement Process Analysis (MPA) Addendum

> If you can measure what you are speaking about, and express it in numbers, you know something about it. But when you cannot express it in numbers your knowledge is … unsatisfactory.
>
> **Lord Kelvin**

Introduction

The benefits of the measurement process analysis (MPA) technology extend beyond being a prelude to performing process capability studies, designed experiments, analysis of variances (ANOVAs), and so forth. This chapter is intended to provide several industry examples demonstrating ancillary benefits that can be derived from management's awareness of the importance of understanding the measurement processes found within the typical manufacturing facility.

Measurement Process Analysis (MPA) and Return on Investment

The management group of any manufacturing facility that has invested resources in educating staff and manufacturing associates in statistical process control (SPC)/Six Sigma technology typically expects a return on investment (ROI). A virtual immediate return on this investment can be realized by conducting analyses on several measurement processes associated with low productivity, poor quality, or excessive scrap. Based on experience in a significant cross section of industry, I estimate the vast majority of measurement processes in use today suffer from one or more of the following shortcomings—instability, lack of discrimination, and lack of reproducibility due to operator technique differences. Consider the potential for good product to be scrapped and bad product to be shipped because an unqualified measurement process is being utilized.

A typical MPA requires four people to measure five samples four times each, consuming roughly one-half of a labor hour. A typical MPA requires absolutely no interference with manufacturing processes. The analysis portion of an MPA, using the Excel spreadsheet available online (www.dougrelyea.com), takes minutes. In less than 1 hour, significant insight can be gained on major problems of low productivity, poor quality, or excessive scrap.

The most effective, least expensive, and least intrusive SPC tool is the measurement process analysis.

MPA Results as a Training Tool

Once a successful MPA has been performed, the control charts, bar graphs, and data resulting from the analysis provide a graphical and numerical representation of stability and discrimination capability of the measurement process as well as differences in operator techniques. Figure 7.1 illustrates the control charts from a measurement system after several significant improvements in operator techniques and local environment had been implemented.

One of Dr. Joseph M. Juran's major contributions to industry was his message on the need to "establish controls to hold the gains." (Juran, 1989, p. 21)* Holding the gains is one of the most difficult aspects of process improvement. So many times problem solvers revisit a problem they once solved only to discover the solutions that were once implemented had been set aside. Time after time, management-implemented solutions based on process capability studies, designed experiments, MPAs, and so forth, fall by the wayside as everyday production pressures prevail and old habits begin to encroach on improvements.

Management should have available to them methods of holding process improvement gains that will enable them to monitor process improvements without detracting from all their other responsibilities.

---

* Juran, J. M. 1989. *Juran on Leadership for Quality*. New York: Free Press.

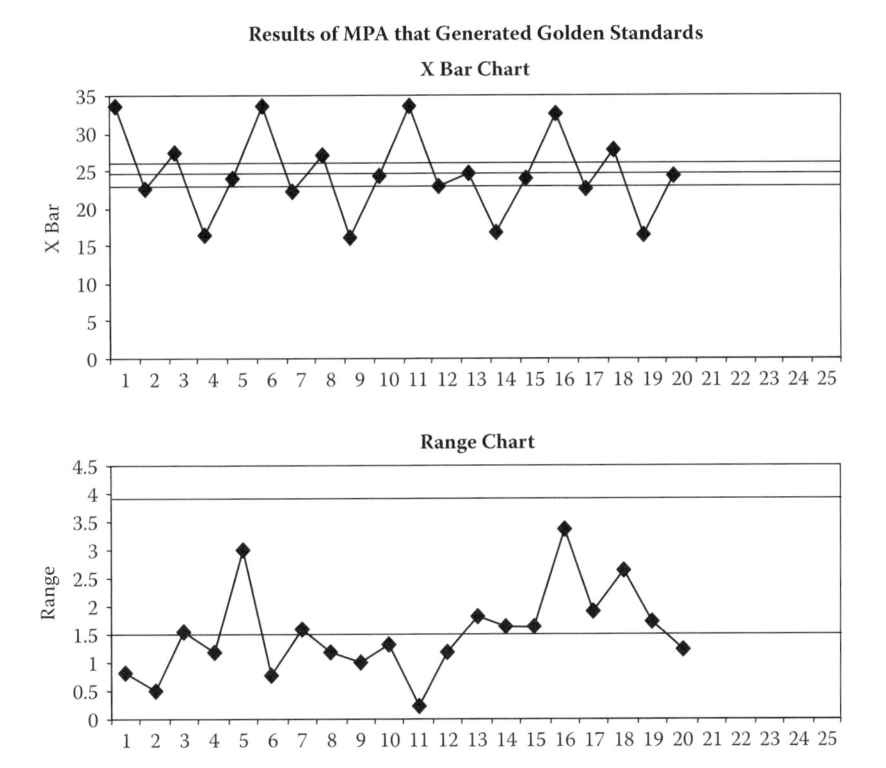

Figure 7.1 Results of measurement process analysis (MPA) that generated golden standards.

One method of holding the gains made by qualifying a measurement process is to use the MPA results to qualify new operators as well as recalibrate experienced operators. We can use Figure 7.1 as an example of holding the gains in the area of measurement processes. The samples that were measured and the two control charts that resulted in Figure 7.1 can now be used to train new operators as well as ensure experienced operators are following the procedures that resulted from the MPAs performed on this measurement process.

Note that the Range chart indicates the measurement process is stable by virtue of all the ranges below the upper control limits. Also, the five averages of each operator on the X Bar chart form virtually identical patterns indicating there are no significant operator technique differences when this measurement system is properly utilized.

The samples used to perform this measurement study should be labeled and stored in a protective container for future reference.

Many times in industry, new operators are instructed in the use of measurement equipment and then allowed to perform measurements and make decisions without any qualification of the effectiveness of their instruction.

The efficacy of a new operator's training in the use of a measurement process could be determined if the samples and results of the MPA for the measurement process in question were available to the supervisor who trained the operator. This effort would merely require the supervisor to follow the same MPA format that was applied to the four original participants but this time only passing the samples through the new operator subsequent to his or her training.

Figure 7.2 represents the X Bar and R control charts from Figure 7.1 with the newly trained operator's pattern of averages and ranges, consisting of

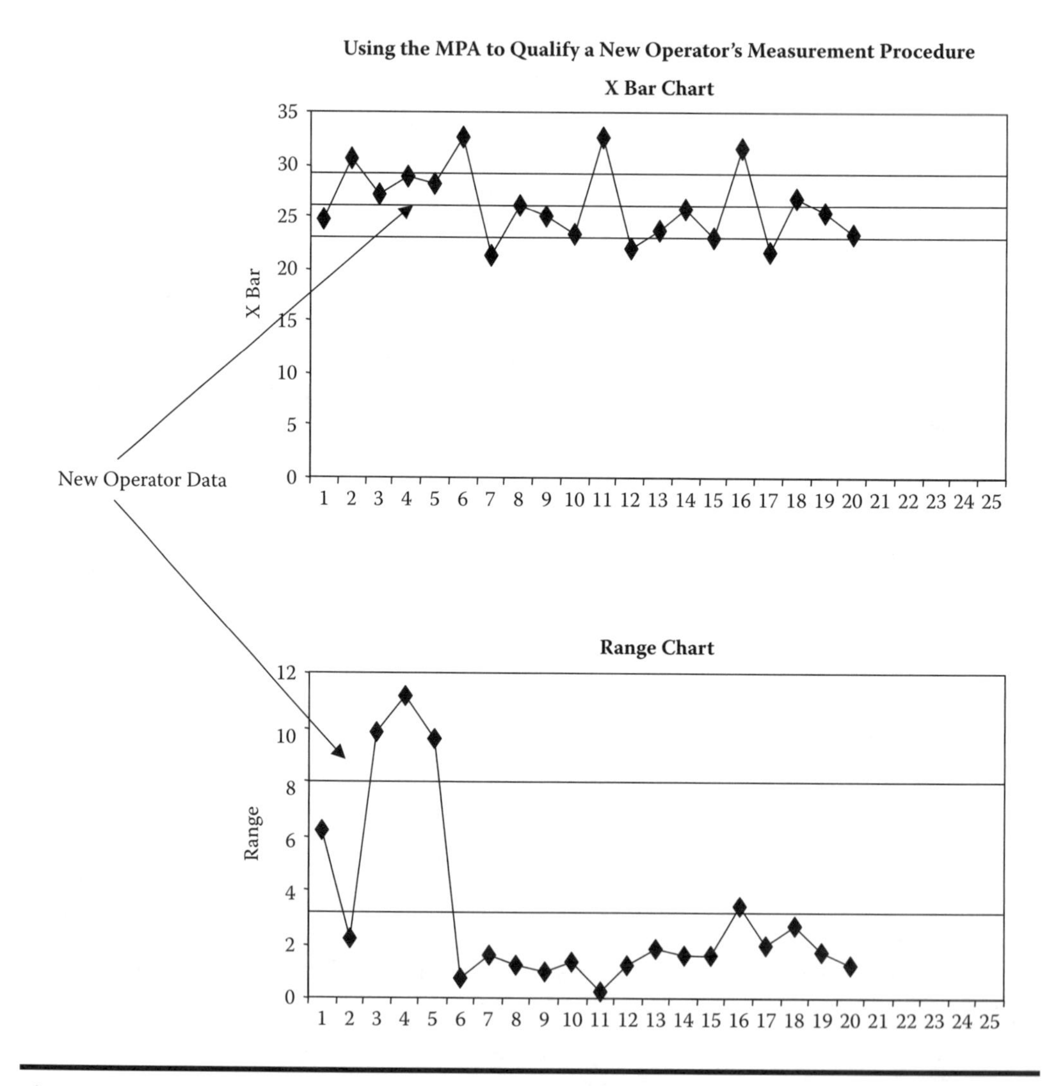

Figure 7.2 Using the measurement process analysis (MPA) to qualify a new operator's measurement procedure.

four repetitions on each of five samples, replacing the data provided by operator 1 on the original MPA study.

The pattern of the new operator is totally different than the pattern of original operators 2 and 3 and 4. Also, when the new operator is part of the measurement process, the measurement process becomes unstable.

The same technique may be used to ensure experienced operators have not adopted technique differences over time.

One major manufacturer safely stores all samples used to perform the final MPA that confirms the measurement process is stable and has a minimal degree of operator technique difference, and employs these "golden standards," as they are called, to evaluate newly trained operators and periodically recalibrate the experienced operators.

MPA as a Measurement Device Purchasing Tool

In Case Study 6.1, "Wire Diameter Micrometer and Microscope Measurement Process Analyses (MPAs)," the wire diameter micrometer measurement process was dominated by error, and it was eventually replaced by the noncontact measurement system of a machinist microscope. To be specific, a machinist microscope was borrowed from a supplier of specialized measurement devices and purchased after it was proven to provide measurement stability, adequate precision, adequate discrimination, and robust performance in the hands of various operators. The measurement device was purchased only after it was demonstrated it was capable of working in concert with the other elements of the measurement process in providing stable and capable measurement results.

In Case Study 6.2, "Elongation Tester Qualification," the manufacturer of polymeric-coated product performed two MPA studies, one on the existing elongation testers and one on a new model of elongation tester that boasted improved resolution. The manufacturer decided against purchasing multiple units of the new model, because the two MPAs indicated both old and new models provided the same degree of imprecision.

In addition to performing an MPA on every measurement process in a manufacturing facility, it is recommended that management mandate that no future measurement equipment intended for analysis of processes is to be purchased until an MPA is performed on a loaner. And if the results indicate instability, significant operator differences, or a discrimination ratio less than four, the measurement equipment should not be purchased.

I would be remiss if I did not offer some clarification when considering the discrimination ratio in the context of deciding to purchase a measurement device.

A manufacturer of electronic components was considering the purchase of a very expensive automated optical measurement device. Before the purchase was finalized, the company was encouraged to perform an MPA at the equipment distributor's location using their own manufactured component for the analysis. The results of the MPA indicated the measurement process was stable with a discrimination ratio of one. There were no operator technique concerns, because the measurement device was automated, and there was no sample preparation involved. The electronic component manufacturing company's management group was dismayed at the failure of the optical measurement device to pass the basic requirements of a measurement process. A discrimination ratio of one is not necessarily a failure of the measurement device—it may be an indication of an extremely well-manufactured product.

Consider the discrimination ratio, in simple terms, is the ratio of the product variation to the measurement process variation. If the normal variation of the measurement process, in terms of sigma, is 0.0002 and the normal variation of the manufactured product characteristic being measured is 0.0002, the discrimination ratio will be one. If the customer's specification for the critical characteristic is ±0.003 and one sigma of the product characteristic variation is 0.0002, the capability index (Cpk) would be equal to 5 when the product average is centered at the customer nominal.

The highest and best use of a measurement process is to understand process variation and to gauge manufacturing process improvements. How much more could a process be improved when one sigma is equal to 0.0002? Why invest in a measurement device that will not help in improving the process?

Measurement Equipment Terminology

On any number of occasions, I have been invited to attend meetings with a client company and a salesperson representing a measurement equipment manufacturer or distributor. The objective of the meeting was always to discuss the precision and discrimination of available measurement equipment my client company was interested in purchasing for purposes of analysis and subsequent process improvement activity.

Questions were typically posed regarding the existence of any data collected on the same or similar product manufactured by my client. Specifically, was there any data from which we could collectively understand how the equipment would interact with my client's product in terms of precision and discrimination?

The answers from the measurement equipment salesperson were invariably couched in terms of accuracy and resolution.

Accuracy, of course, is comparison to a known standard. Accuracy is a necessary characteristic of any measurement equipment, but it has nothing to do with the ability of the measurement equipment to measure product with the necessary precision (repeatability) to support manufacturing process improvement.

Resolution is also important. Resolution is the smallest measured unit the measurement equipment can record. A digital micrometer that has a resolution of 0.001 inch cannot measure to 0.0015, such a micrometer has the ability to measure a part in increments of 0.001 inch, and no less. Improved resolution is also desirable when measuring a critical characteristic on manufactured product. But there is no benefit to a manufacturer if purchased measurement equipment has a resolution to 0.0005 if the product characteristic being measured evidences normal variation, in terms of sigma, equal to 0.0001.

It would be impractical for a measurement equipment manufacturer to perform a proper MPA using every possible material configuration that potential users might use in their manufacturing process. However, it is very important for measurement equipment companies, as well as potential buyers of the equipment, to understand that accuracy is not precision and resolution is not discrimination. Manufacturers and distributors of measurement equipment often promote a new, improved device based on increased resolution as in Case Study 6.2, "Elongation Tester Qualification." And in Case Study 6.2, it was proven that there was no significant difference in terms of discrimination between the new and old models in spite of the increased resolution of the new model.

Measurement Variation and Specification

Some customers have a requirement that the amount of measurement variation must not exceed 10% of the critical product characteristic involved in an MPA. It would certainly be appropriate to satisfy such a customer requirement that is aimed at having a level of confidence that the measurement process can identify product that is out of specification. However, it is important to realize that such a requirement provides no assurance the measurement process can be employed to tell the difference between samples made during periods of normal variation, which is basic to improving the manufacturing process. It is quite possible that a measurement system with a discrimination ratio of one would still qualify as being useful for measuring a specific customer's product based on that customer's specification. Figure 7.3 (not drawn to scale) illustrates this concept.

Although a measurement process may be acceptable to a customer if the measurement variation is equal to or less than 10% of specification, that same measurement process might not be appropriate for performing process capability studies, ANOVAs, and so forth.

At the very least, when performing an MPA to satisfy a customer requirement that the measurement process be compatible with customer specification, it is recommended that the discrimination ratio calculation be performed as well. Knowledge of the discrimination ratio can only assist the manufacturer to better understand the full capability of the measurement process.

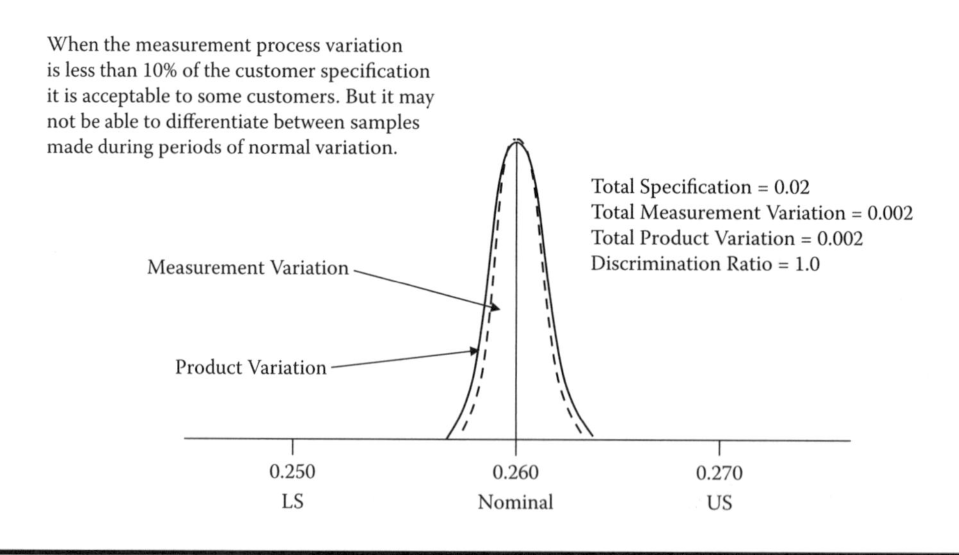

Figure 7.3 When the measurement process variation is less than 10% of the customer specification, it is acceptable to some customers. But it may not be able to differentiate between samples made during periods of normal variation.

Automated Measurement and Feedback Systems

An American company engaged in extruding thin layers of polymeric material onto substrates was approached by a manufacturer of automated measurement devices designed to sense the thickness of extruded coatings and automatically adjust die openings to maintain a consistent coating thickness with minimal variation. The equipment supplier proposed the $350,000 investment for this equipment would be returned within 18 months as a result of the promised reduction in the thickness variation of the extruded material.

The measurement equipment manufacturer based this return on investment calculation on coating thickness data provided by the extrusion company. The measurement company had requested and received coating thickness measurements that had been collected by extruder operators during their shifts. The existing method of measurement required the operators to weigh a sample of the substrate being processed, and then weigh a sample of extruded product from the end of a roll. The extruder operators utilized the difference between the two sample weights to calculate the coating thickness. Data sets from approximately nine shifts spanning a period of 2 weeks were forwarded to the equipment manufacturer.

The data sent to the equipment manufacturer appear in Table 7.1.

The measurement equipment manufacturer analyzed the data and provided the extrusion company with results in terms of average and standard deviation of coating thickness.

The average was calculated to be 0.0016, and one standard deviation was calculated to be 0.00015. The measurement equipment manufacturer stated the proposed automated measurement feedback system would reduce the standard deviation by a factor of three, which was the basis of the estimated return on investment.

I received the data while in Europe with a request to confirm the calculations for average and standard deviation. My calculations agreed with the average, but my standard deviation was 0.00003, approximately 20% of the standard deviation suggested by the equipment company.

Why the Huge Difference?

Nobody was being dishonest. The equipment measurement company used the classic calculation for standard deviation as provided by many software packages, including Excel. This calculation does not take into account

Table 7.1 Coating Thickness Measurements Reported to the Measurement Equipment Company

0.001494	0.001289	0.001589	0.001571
0.001474	0.001417	0.001589	0.001604
0.001505	0.001411	0.001204	0.001606
0.001526	0.001386	0.001309	0.001628
0.001524	0.001373	0.001303	
0.001535	0.001392	0.001333	
00.01456	0.001396	0.001312	
0.001495	0.001397	0.001309	
0.001522	0.001359	0.001334	
0.001478	0.001414	0.001296	
0.001486	0.001713	0.001304	
0.001683	0.001751	0.00151	
0.001703	0.001712	0.001502	
0.001711	0.001696	0.001483	
0.001673	0.001679	0.001502	
0.001694	0.001698	0.001476	
0.001686	0.001706	0.00151	
0.001669	0.001728	0.001498	
0.001308	0.001682	0.001611	
0.001279	0.00173	0.001652	
0.001282	0.001682	0.001599	
0.001276	0.001572	0.001614	
0.001305	0.00161	0.001624	
0.001306	0.00158	0.001611	
0.001327	0.00161	0.001606	

process average shifts. The standard deviation I provided was based on the average moving range method as discussed in Chapter 5.

Figure 7.4 illustrates the coating thickness data plotted on an Individuals control chart. The changes in process averages spanning the nine shifts are apparent. Further investigation confirmed initial suspicions that the process average shifts were a result of differences in operator setup techniques.

Management decided to concentrate on making the setup procedures more consistent and forego purchasing the automated equipment, which, in this case, would only be compensating for a lack of operator training.

I would be remiss if I did not point out that neither the operator's manual technique of determining the coating thickness nor the automated equipment being promoted by the measurement equipment company, at the time of the analysis, had ever been subjected to a proper MPA. When the manual measurement method was eventually performed, it was discovered to have a discrimination ratio of one. Although the manual method could be used to set the process average, it could not be used to improve the process, and if all the operators used the same procedure for setting the process average and refrained from making unnecessary adjustments, the batch-to-batch coating thickness variation would be minimized.

Understanding the measurement process, and the terminology and analysis methods of measurement equipment manufacturers and suppliers, can provide any number of competitive benefits.

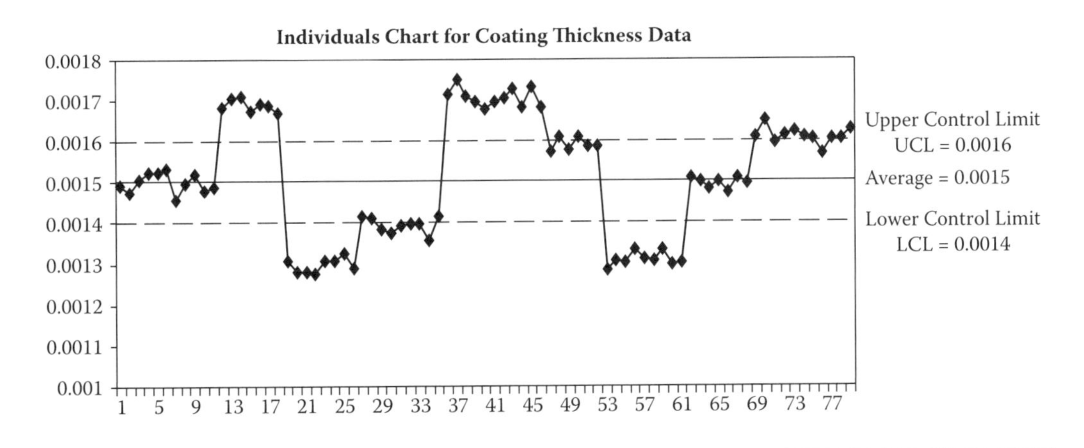

Figure 7.4 Individuals chart for coating thickness data.

Manufacturing process control comes about by controlling the elements that make up every manufacturing process—people, measurement process, raw material, equipment, local environment, and methods. The easiest process element to control and the one that generally offers the quickest and most significant manufacturing process improvement is the measurement process.

Chapter 8

Creating Process Control Charts

Science is the great antidote to the poison of enthusiasm and superstition.

Adam Smith
The Wealth of Nations, 1776

Overview

When a process capability study is completed and a customer is informed the process is stable and capable with a capability index (Cpk) equal to or exceeding the customer's requirements, there is a tendency, on the part of many management groups, to be satisfied with the process at this juncture. Future activity on such a process is usually limited to the maintenance of statistical product control charts to ensure incidents of instability are compensated for by means of adjusting process parameters.

As a result, statistical product control charts are found on manufacturing shop floors far more often than statistical process control charts. Several root causes contribute to the paucity of statistical process control charts in manufacturing. Limited engineering resources are likely the primary reason organizations do not evolve to process control. New product development and pressing problems of scrap and rework on other product lines prevent technical staff from refining those processes that are already performing satisfactorily. Also, there is the very real fact that many people generally feel more comfortable making process decisions based on product measurements as opposed to input from key process parameters.

As a practical matter, evolving a stable and capable process from product control to process control does not necessarily require engineering resources. A shop floor associate educated in the concepts of product control and process control, while operating the equipment and updating the statistical product control chart, can support the creation of an Individuals statistical process control chart with very little external support.

Over time, operators, supervisors, and engineers tend to identify key process parameters in a rather informal joint effort. If, while monitoring a statistical product control chart, an operator detects a violation of the rules of instability, it is the responsibility of that operator to make a note on the product control chart as to what he or she believes is the cause of the instability and to compensate, if possible, by adjusting the process parameter generally accepted as being key to the stability of the process. If the operator cannot identify the cause of the instability or the adjustment is not reflected on the product control chart as having corrected the problem, then the supervisor and engineers need to be notified. A discipline of utilizing the product control chart to seek and identify the process parameter causing a particular form of instability will invariably be successful. Once the key process parameter has been factually identified and corrected (equipment components repaired, modified, replaced, etc.), an Individuals process control chart should be constructed to eventually replace the product control chart.

The first goal when establishing the initial process control chart is to identify the amount of normal variation of the process parameter of interest.

The supervisor can create the initial process control chart with input from operators accompanied by a final review and approval of engineering. The process control chart can begin with a centerline, which could represent the standard operating procedure (SOP) required setting for the identified key process parameter. Temporary upper and lower "control limits" could be established with operator input.

For example, if analysis has indicated zone A temperature for an injection molding process to be a key process parameter contributing to scrap for dimensional characteristics, then zone A temperature controller should be inspected for operational integrity. An Individuals statistical process control chart should be constructed for the process parameter zone A temperature. The chart could be constructed in the following manner. If the SOP requires zone A temperature to be set at 450°F, the centerline of the process control chart that is being initiated could be chosen to agree with the SOP requirement (450°F). If the shop floor associates have, over time, set zone A temperature as high as 460°F and as low as 440°F when trying to eliminate dimensional discrepancies, then the process control chart upper and lower temporary limits could be established at 440°F and 460°F, respectively. Prudence would dictate this decision receive the blessing of the molding engineers.

If a designed experiment, analysis of variance (ANOVA), or the results of a process capability study indicate a centerline temperature that is more desirable than the SOP, perhaps that centerline setting could be chosen and bracketed with agreed on temporary control limits. The important aspect of this concept is to begin the process control chart with a centerline indicating the temperature at which zone A should be set at the beginning of the production run, and to provide upper and lower limits (simulating plus and minus three sigma limits) that zone A temperature should not exceed. Using dotted lines, the distance between the temporary limits and the centerline can be divided into three segments either side of the centerline to simulate the one- and two-sigma limits that would normally be part of an Individuals control chart with properly calculated limits.

Each time the operator measures the key critical product characteristic, which is of concern, and records the results on the product control chart, the operator should, at that time, also record the temperature reading for zone A on the process control chart. From this point on, the process control chart should be monitored just as if it were a product control chart. The operator should make a note if instability is detected on either the product or process control charts. If the process parameter being studied is truly affecting the product dimension, then instability on the product control chart and the process control chart should be coincident. But indications of instability on either chart should now require attention to be directed at finding the root cause of the process parameter—zone A temperature—instability. Voltage fluctuations due to energizing or deenergizing equipment might be one possible root cause in such a situation.

It is fully recognized that at the outset of developing a process control chart, the normal variation of zone A temperature might very well exceed the established temporary control limits. Therefore, when the corresponding product measurement is within the calculated control limits on the product control chart, it would not be prudent to make adjustments to the zone A temperature controller.

When a minimum of 30 data points have been recorded on the temporary process control chart, the recorded data can then be used to calculate plus and minus one-, two-, and three-sigma limits to replace the temporary control limits for the Individuals process control chart.

A process control chart can be created within as little as 2 days without direct involvement of the engineering staff.

CASE STUDY 8.1 SLUG WEIGHT PRODUCT CONTROL TO HEAD PRESSURE PROCESS CONTROL

A manufacturer of engineered rubber products had traced a serious defect that was affecting virtually dozens of part numbers back to an interim extrusion process that was common to an entire product line.

Process capability studies had been performed on the entire product line, by part number, for the key product characteristic "slug weight" long before the current problem had reached its current, serious stage. Each process capability study had generated a statistical product control chart for each of the extruded part numbers studied, resulting in dozens of statistical product control charts—one for each product. Each time the extruder was set up to accommodate a particular part number, an X Bar and R product control chart with preprinted centerlines and control chart limits specific to that part number was posted at the extruder. Operators were required to select five samples during a 30-minute period, weigh each sample on a gram scale, calculate the average and range of the subgroup of five samples, and record the results on the respective X Bar and R charts.

When the root cause of the product line difficulty was traced back to the extruder, the problem solvers first went to the results of the extruder process capability study for the product currently being manufactured. Figure 8.1 illustrates the X Bar and R charts for this study. Based on the two charts of Figure 8.1, it was obvious the process had demonstrated stability during the study.

Next the problem solvers went to the volumes of completed statistical product control charts that were on file to see if the extruder had encountered instability that would help to explain the current problem. Many of the completed charts demonstrated violations of X Bar and R chart instability detection rules, but in each incident of instability, the only notes made by the operators were "made adjustments." There was no indication that any attempt had been made to identify the process parameter that might have caused the instability. There was no indication as to which process parameter or parameters the operator had adjusted to compensate for the instability. Unfortunately, this improper use of statistical product control charts in industry today is the rule rather than the exception.

For the next several days, an engineering technician on each of the three shifts assumed responsibility for selecting samples from the extruder, weighing the samples, calculating the average and range of each data set of five samples, and updating the X Bar and R product control charts. The

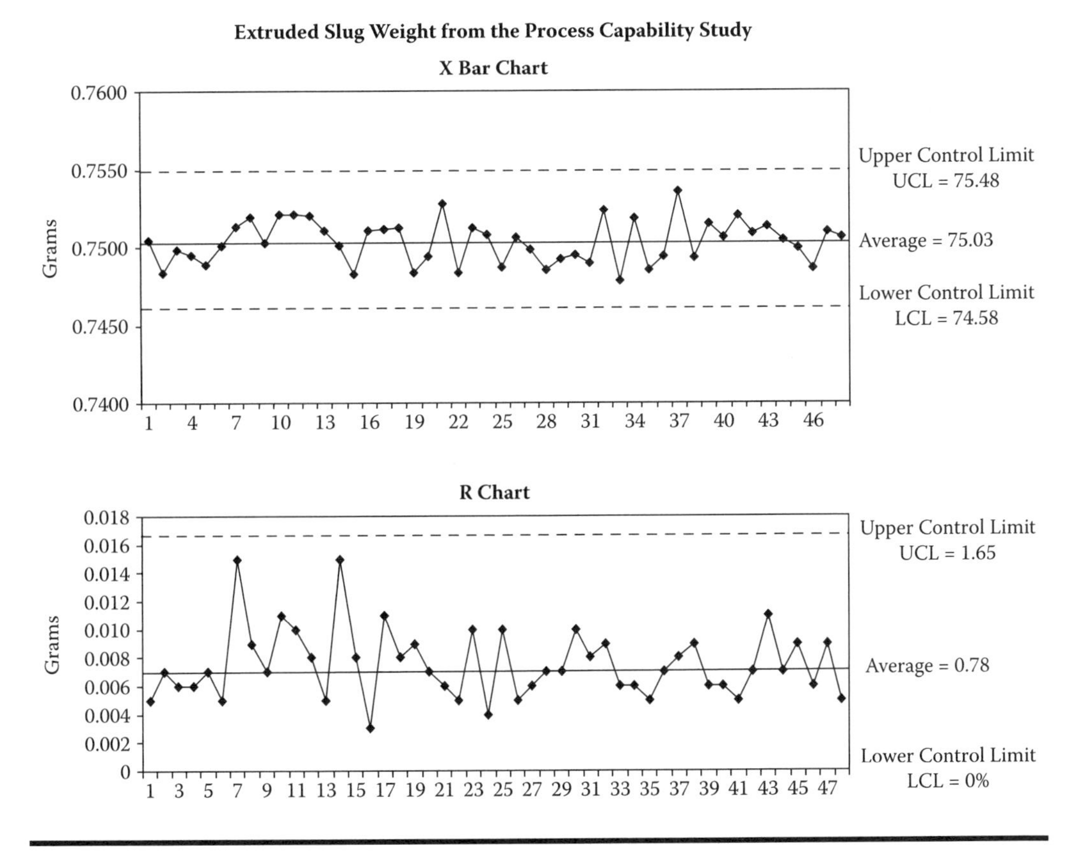

Figure 8.1 Extruded slug weight from the process capability study.

technicians were instructed to collect data, record the results, and observe only. An indication of instability appeared on the X Bar chart several hours into the second shift of the technicians' involvement followed by two more incidents of instability. See Figure 8.2, points 1, 2, and 3. The operator chose to make no adjustment for the first incident (one data point below the lower control limit). But the operator did make adjustment to the head temperature for incident number 2 (data points below the lower limit as well as eight consecutive data points on one side of the centerline, and another set of adjustments for incident number 3 (again, data points below the lower control limit as well as eight data points below the centerline). More importantly, the technician noticed each incident of instability was accompanied by higher extruder head pressure as indicated on the head pressure gauge.

At the end of the shift, the extruder operator made a note in the logbook and verbally advised the third-shift operator that the screen pack should be changed and the breaker plate cleaned.

The screen pack is that element in the extruder that screens out foreign matter and undispersed chemicals before the raw material reaches the extruder die. The breaker plate is a metal plate that provides support to

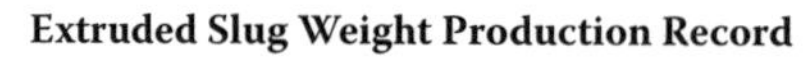

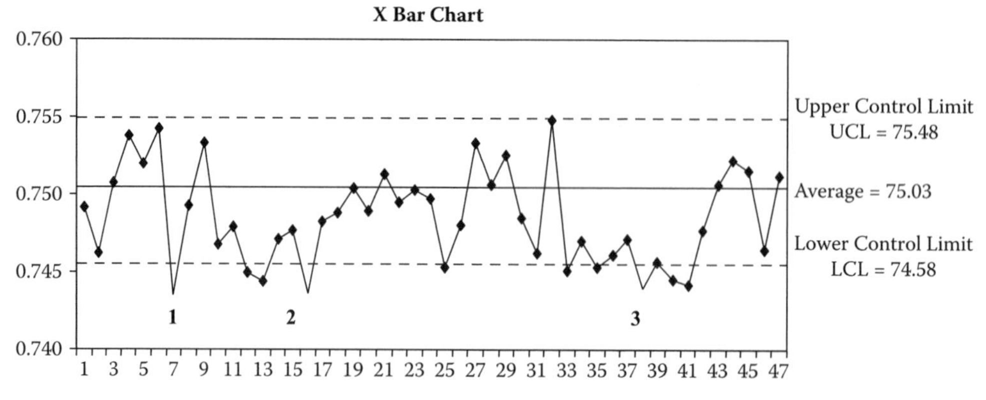

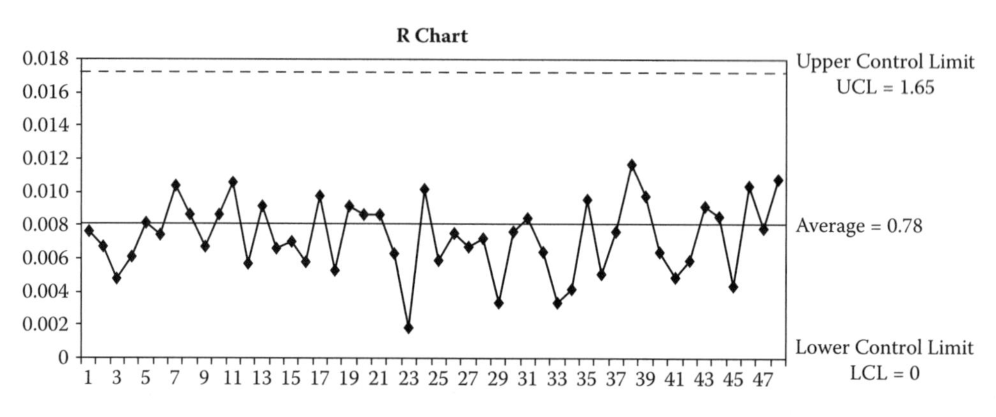

Figure 8.2 Extruded slug weight production record.

the screen pack, which is under constant pressure from the material being extruded. Changing the screen pack and cleaning out the breaker plate are undesirable jobs that are generally avoided by the extruder operators.

At an early meeting the next morning, the group decided there was a high probability the screen pack was the process parameter that was contributing to the weight instability, and the head pressure was the measurable process parameter that could indicate when the screen pack should be changed. It was decided to construct a statistical process control chart for the process parameter head pressure.

A meeting was held with production supervisors and extruder operators to review the discoveries, reinforce the proper use of statistical product control charts, and introduce the plan to construct a process control chart. An Individuals process control chart was introduced with preprinted centerline and temporary upper and lower control limits. It was explained that the centerline was the head pressure the engineering department would expect to be generated for a specific process combination of die, screw speed,

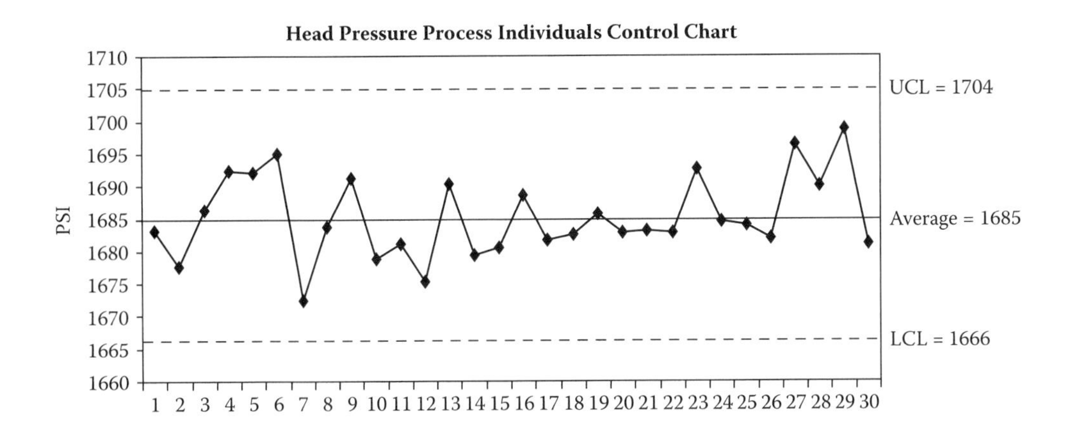

Figure 8.3 Head pressure process individuals control chart.

screen pack, and so forth. The upper and lower limits represented the variation of head pressure that was estimated to be normal to this process.

Operators were instructed to record the head pressure on the Individuals process control chart each time they recorded product data on the X Bar and R product control chart.

The product control chart and the temporary process control chart were reviewed often by production supervisors over the next several days to ensure operators were properly recording data on both charts, following all the rules for detection of instability, and, more importantly, changing over the screen pack at the first sign of instability on either the product control chart or the process control chart. After several weeks, the data from the process control chart were used to calculate valid upper and lower control limits for the process control chart.

Figure 8.3 illustrates the statistical process control chart with properly calculated upper and lower control limits for extruder head pressure. The plan going forward was to require the X Bar and R product control chart and the head pressure process control chart to be maintained in tandem and to change the screen pack at the first sign of instability on the process control chart. This would continue until management and the operators were comfortable in reducing product measurements to one sample every half hour and use an Individuals product control chart in lieu of the more sample-intense X Bar and R chart combination. As comfort levels rose with using the head pressure process control chart as an indicator of process instability, it was hoped that product samples could be reduced to one per hour for recording on the Individuals product control chart. Of course. the goal was to eventually eliminate product sampling and measuring for purposes of product control in favor of relying on process control based on head pressure variation. This usually takes a long time.

Chapter 9

Process Capability Studies beyond the Shop Floor

> Be precise. A lack of precision is dangerous when the margin of error is small.
>
> **Donald Rumsfeld**

Introduction

It is difficult to create a culture that requires engineers and shop floor associates to use industrial statistics such as process capability studies, design of experiments, analysis of variance (ANOVA), and so forth, to solve shop floor problems and otherwise improve production processes. The only effective way to create such a culture is for management, beginning at the highest level, when presented with a problem to ask the prime question—"What kind of problem is it, is it one of instability or is it one of incapability?"

Management also needs to look beyond the shop floor when considering the advantages of the process capability study.

Research and Development

Many manufacturing shop floor problems begin at the end of the research and development (R&D) process. In many organizations, a new process is released to production based on a prototype process under constant attention of engineers and technicians successfully producing a small quantity of product that meets customer specifications. Seldom is a new process introduced to production accompanied by documentation indicating the new process demonstrated stability and capability in the R&D environment.

A standard operating procedure (SOP) for transitioning a new process from R&D to production must include a requirement for data and the appropriate control charts, indicating the R&D department performed process capability studies for each of the critical product characteristics. The studies, performed with qualified measurement processes, must demonstrate each critical product characteristic was proven to be stable and capable while produced under R&D auspices.

When the process has been established on the production floor and the production manager is satisfied with the R&D documentation, the SOP should then require a process capability study be performed on the shop floor involving floor associates. It should not be assumed that a process that is stable and capable in the R&D laboratory will also be stable and capable on the manufacturing shop floor.

If the production process capability study meets the organization's requirements for stability and capability (capability index [Cpk] = 1.33 minimum), the new process should then become the responsibility of the production department. In the absence of a successful process capability study performed under shop floor conditions, the process should remain the responsibility of R&D or engineering until the process can be corrected.

Sometimes market conditions and customer requirements demand that a new process that is unstable or incapable produce product at maximum capacity while R&D or engineering identify and correct whatever problem or problems exist. If management decides to permit a process that is unstable or incapable to continue producing product, the decision is made with a full understanding of the cost associated with scrap, rework, and potential customer returns. Equally important is the fact that management understands the problem is not due to some shortfall on the part of the production staff but rather needs to be solved by means of engineering efforts.

Under the best of conditions, it is often difficult for the production department staff and the shop floor associates to ramp up a new process to full production. When an unstable or incapable process is released from

R&D to the shop floor, any potential difficulties are increased an order of magnitude. In addition to scrap, rework, and quality problems, barriers between R&D and manufacturing also tend to be created.

Front ending the release of a new process with an R&D process capability study can serve to eliminate many of the problems typically associated with introducing a new process to full production status.

Sales and Marketing

Many sales and marketing professionals in industry today are familiar with the concepts of continuous improvement, and when speaking with existing or potential customers speak in amorphous terms such as "we aim to exceed our customer's expectations." Exceeding a customer's expectations is an excellent goal; however, it would be more effective if the goal were more clearly defined.

Sales and marketing professionals need to be thoroughly conversant with the concepts of stability and capability and be comfortable with offering a customer results based on process capability studies performed on the customer's product. It would be constructive to point out to a customer that the customer's Cpk requirement of 1.33 is being exceeded by the supplier's capability to provide a Cpk of 1.5 on all critical product characteristics. Of course, this will translate into more parts being produced closer to nominal, which should result in quicker assembly of components at the customer's site, and thus reducing labor costs. Over time, as the plan to increase the Cpk develops, perhaps the customer could further reduce cost by eliminating incoming inspection, and so on.

It can only further a supplier's interest if sales and marketing professionals are speaking in terms of producing more product closer to the customer's nominal while the supplier's competition is still speaking in terms of producing all product within specification.

Purchasing

Purchasing professionals can have an incredible impact on the competitiveness of a manufacturing facility if they have a full understanding of the concepts of stability and capability.

In my experience, the two major causes of instability in manufacturing are human interference (operator adjustments) and raw material variation between batches. Oftentimes, the first is caused by the second.

In some manufacturing facilities, the operators have become so accustomed to changes in the process output when a different batch of material is introduced that they automatically begin to make adjustments anytime one batch runs out and a new one is picked up, regardless of the batch number.

It would be appropriate for management to charge purchasing professionals with responsibility to require suppliers to provide results of process capability studies for product characteristics that are important to internal manufacturing processes. At some point in time, the purchasing management, with engineering support, should develop a minimum Cpk requirement for all purchased raw material critical characteristics.

Shop Floor Practices

It is not uncommon in manufacturing facilities that do not have a culture of using process capability studies, and discussing processes in terms of stability and capability, for various operators, based solely on preference, to produce product from one end of the specification to the other. Some operators prefer to run close to the low end, and other operators prefer to run close to the high end of specification. There are even a few operators who intuitively aim for the customer nominal. Consider the effect of this practice when the different outputs produced at different averages reach the customer's process.

A brand-name manufacturer of mattresses was finding it necessary to make time-consuming and costly adjustments to the automatic upholstery equipment that applied and stitched fabric over the bare spring mattress units received from a supplier. An investigation showed the received units ranged from the very low end to the very high end of the customer's length specification. A review of the equipment at the supplier's facility discovered several process streams producing product at the low end of specification with little variation and several other process streams producing product at the high end of specification with little variation. Corrections were quickly effected by centering all the process streams at the nominal specification, and the customer's problem was solved.

This type of problem should never have existed, because it stems from a culture that is specification oriented as opposed to variation oriented. A variation-oriented culture comes about over time and is created, in part, by involving shop floor associates and supervisors in the concepts, execution, and bottom-line analysis of process capability studies. Also, it is incumbent upon management to force the use of process capability studies by asking the prime question—"What kind of problem is it? Is it a problem of instability or incapability?"

In Conclusion

My fervent hope is that the reader now has a full understanding of how and when to make the highest and best use of process capability studies. Above all, avoid wrapping the process capability study in a blanket of bureaucracy only to be unfolded and used by a few elite individuals in quality or engineering. Strive to make everyone in the organization, from the lead operators to the sales and purchasing managers, fully conversant with the concepts of the process capability study and the results in terms of stability and capability. Make certain your suppliers comply with your Cpk standard, and ensure your customers understand the benefits of dealing with a manufacturer that is striving to exceed their expectations with science and measurable metrics.

Good luck.

Appendix

Formulas and Factors

Chapter 2

Individuals control chart calculation: The control chart illustrated in Figure 2.3 is an Individuals control chart based on the average moving range ($\overline{MR}$).

Average, X Bar ($\bar{X}$) = sum of all 60 data points divided by the number of data points.

Moving range (MR) = the absolute difference (no plus or minus) of two consecutive data points. MR1 = the absolute value of data point one minus data point two, MR2 = the absolute difference of data point two minus data point three, MR3 = the absolute value of data point three minus data point four, and so forth, for a total of 59 moving ranges.

Average moving range (MR) = the sum of the 59 moving ranges divided by 59.

Upper control limit (UCL) and lower control limit (LCL) for the Individuals control chart:

UCL = Average + [(Average Moving Range/d_2) × 3]

LCL = Average – [(Average Moving Range/d_2) × 3]

The d_2 values can be found in Table A.1. The value for a subgroup size of two is selected because each moving range was developed using two values.

Instability detection rules for the Individuals control chart:

1. One data point outside either three-sigma limit
2. Three consecutive data points on the same side of the centerline, and two of the three data points outside the same two-sigma limit

3. Five consecutive data points on the same side of the centerline, and four of the data points outside the same one-sigma limit
4. Eight consecutive data points on the same side of the centerline

Six Sigma:

Six Sigma requires suppliers to maintain a capability ratio of 2 on critical product characteristics when the product average is centered at nominal.

For example,

If customer specification = 0.500 ± 0.015 – total specification = 0.030.

If one sigma were equal to 0.005, three sigma either side of center (six sigma) = total variation = 0.005 × 6 = 0.030.

CR = 030 ÷ 0.030

CR = 1

If, on the other hand, one sigma were equal to 0.0025, three sigma either side of center (six sigma) – total variation = 0.015.

CR = 0.030 ÷ 0.015

CR = 2

When the product average is centered on nominal and one sigma of a product characteristic is multiplied by six, and that full range of variation consumes no more than 50% of the specification, the product is *Six Sigma* qualified.

Chapter 3

X Bar and $\bar{X}$ R Chart Calculations

$\bar{X}$ (Average) = X total (sum of data points in subgroup) divided by number of data points in subgroup.

R (Range) = Largest measurement in subgroup minus smallest measurement in subgroup.

$\bar{\bar{X}}$ Grand Average = Summation of all averages divided by number of averages.

Upper and Lower Control Limits (UCL and LCL) for X Bar and R chart:

$$\text{UCL for } \bar{X} = \bar{\bar{X}} + (A_2 \times \bar{R})$$

$$\text{LCL for } \bar{X} = \bar{\bar{X}} - (A_2 \times \bar{R})$$

$$\text{UCL for } R = \bar{R} \times D_4$$

$$\text{LCL for } R = \bar{R} \times D_3$$

Instability Detection Rules for the X Bar Chart

1. One data point outside either three-sigma limit
2. Three consecutive data points on the same side of the centerline and two of the three data points outside the same two-sigma limit
3. Five consecutive data points on the same side of the centerline and four of the data points outside the same one-sigma limit
4. Eight consecutive data points on the same side of the centerline

Instability Detection Rules for the Range Chart

1. One data point outside the three-sigma limit
2. Eight consecutive data points on one side of the centerline

Chapter 4

Percent Defective (P) Chart Calculation

Average Percent Defective (P) = total (sum of percent defective results) divided by number of results.

$$\text{UCL} = \bar{P} + 3\sqrt{\frac{\bar{P}(1-\bar{P})}{n}}$$

n = sample size

$$\text{LCL} = \bar{P} - 3\sqrt{\frac{\bar{P}(1-\bar{P})}{N}}$$

Instability Detection Rules for the P Chart

1. One data point outside the three-sigma limit
2. Eight consecutive data points on one side of the centerline

Chapter 5

Calculated Method of Standard Deviation

$$\text{One calculated standard deviation} = \sqrt{\frac{(\bar{X}-X_1)^2+(\bar{X}-X_2)^2+(\bar{X}-X_3)^2}{n-1}}$$

etc.

Table A.1 Control Chart Factors

Number of Samples in Subgroup (n)	A_2	D_3	D_4	D_2
2	1.880	0	3.267	1.128
3	1.023	0	2.574	1.693
4	0.729	0	2.282	2.059
5	0.577	0	2.114	2.326

$\bar{x}$ is the average of all the data points, $x_1 + x_2 + x_3 \ldots . \div n$
x_1 is data point one, x_2 is data point two, x_3 is data point three, and so forth
n = number of data point

Estimated Method of Standard Deviation

One estimated standard deviation = $\overline{\text{MR}} \div \text{d}_2$.

Moving range (MR) is the absolute difference (no plus or minus) of two consecutive ranges. For example, MR_1 is the absolute difference of data point 1 minus data point 2; MR_2 is the absolute difference of data point 2 minus data point 3; MR_3 is the absolute difference of data point 3 minus data point 4; and so on.

Average moving range (MR) is the total of all moving ranges divided by the number of moving ranges.

d_2 is a constant from Appendix Table A.1. For a subgroup size of 2 (each MR is the difference of *two* ranges), $\text{d}_2 = 1.128$.

Glossary

Accuracy: Comparison to a known standard. A measurement device that measures a known standard without appreciable error is said to be *accurate.*

Assignable cause: An unusual occurrence that takes place within a process: unusual in that the variation resulting from an assignable cause is outside normal variation.

Attribute data: Counts or quantitative data that have only two conditions (good or bad, acceptable, rejectable, etc.). *Attribute data* are whole numbers: for example, number of rejects.

Average (X Bar): The mean value of a subgroup. The average is calculated by adding the values within the subgroup and dividing by the number of values.

Brainstorming: An exercise in which a group of individuals express and record ideas concerning a specific problem. The essence of such an exercise lies in the openness and creativity of a free and unbiased exchange of ideas.

Capability: The ability of a process to produce all product well within customer specification. A process is said to be capable when it meets the minimum requirement of having the process average at nominal and taking up no more than a customer-specified amount of the specification.

Capability index: The numerical value applied to the relationship between the total variation of a product characteristic and the customer specification. A capability index (Cpk) of 1.33 is the minimum requirement of many manufacturers.

Cause and effect exercise: A problem-solving tool used to identify the root cause of a problem. Fishbone diagrams and brainstorming are very effective means used to identify the root cause.

Control chart: A statistical process control tool that initially helps define the stability and capability of a process. The initial result of a process capability study is a *product control chart* that should evolve into a *process control chart.*

Control limits: Boundaries on control charts which identify the amount of normal variation we can expect from a product or process.

Cpk: See Capability index.

Discrimination: The ability of a measurement process to detect samples made during periods of normal variation.

Discrimination ratio: A numerical value approximately equal to the amount of product normal variation in terms of sigma divided by the amount of measurement process variation in terms of sigma. A discrimination ratio equal to or greater than 4 is required for a measurement process to be qualified for purposes of statistical analysis of a process output.

Histogram: A graphic tool that illustrates, in bar chart form, the distribution of data.

Individual control chart: A form of control chart that uses the range between individual data points to develop the control chart limits. An individual control chart is very useful in the analysis of continuous processes such as extrusion and destruct measurement processes. The control chart limits of an Individuals chart, unlike the control chart limits of the Averages (X Bar) chart, also relate directly to the total amount of normal variation of the characteristic being studied.

Instability: The state that results in a process when assignable causes are present.

Lower control limit: The calculated lower boundary of a control chart representing the minus three sigma limit of the normal distribution.

Manufacturing process: Any combination of people, a measurement process, raw material, local environment, and standard procedures formed to complete an assigned task of manufacturing or service. Machining is a process, as is receiving and executing purchase orders.

Measurement process analysis (MPA): The statistical study of a measurement process to determine if it is precise and has an adequate *discrimination ratio.*

Moving range (MR): When analyzing a set of data, the moving range of two consecutive data values is the absolute difference of those two values. For instance, MR1 is the absolute difference of data point 1 minus data point 2 (absolute value implies no plus or minus is assigned to the resultant MR), MR2 is the absolute difference of data point 2 minus data point 3, and so on. For 50 data points, there would be 49 moving ranges.

Nominal: The midpoint of a specification. In general, the nominal specification is the most desirable to achieve.

Normal curve: Sometimes called the bell curve because of its shape. The *normal curve* is characterized by having one peak and symmetrically trailing off to extreme ends on either side of the middle. A *normal curve* results when variable data are plotted from a process operating without any assignable causes present—only normal variation.

Normal variation: Variation due to natural and random occurrences. The differences of height among high-school seniors are a good example of *normal variation.*

Pareto chart: A tool that displays in bar chart form a frequency of occurrence within certain categories of interest. Usually the highest category appears in the extreme left, and subsequent categories are presented in descending order. A Pareto chart separates the vital few categories from the trivial many.

P chart: A type of attribute chart that indicates when the percent defective product yielded by a process has been adversely affected by an assignable cause.

Percent defective: The percentage of a population that does not conform to specification.

Precision: The ability of a measurement process to record the same value within the bounds of normal variation when a measurement is repeated on the same sample, in the same location, by the same person, using the same measurement device.

Process average: The central tendency of a process. The process average is determined by collecting and converting data to a form such as a histogram or control chart.

Process capability study: The practice of analyzing product variation in order to determine if the process is stable and is capable of producing product that meets the customer's minimum Cpk requirement.

Process parameter: Usually a controllable factor related to one of the process elements, such as equipment line speed or raw material melt index.

Range: The difference between the largest and the smallest value with a group of variable data.

Sample: Statistical process control is based on the concept that the quality of a entire process can be confidently determined through the statistical evaluation of groups of random samples.

Sigma: See Standard deviation.

Stability: The ability of a process to produce consistent and predictable output. A stable process has only normal variation present; no assignable cause variation is present.

Standard deviation: A means of measuring the width of a distribution. A normal curve has three *standard deviations* either side of the process average. The term *sigma* is often used interchangeably with *standard deviation.*

Statistical process control (SPC): A term used to describe the tools and techniques that help to identify the sources of and reduce process variation.

SPC: See Statistical process control.

Statistical product control: A term used by the author to describe a misuse of statistical process control. When statistical process control charts are used to monitor product and make adjustments to the process to compensate for periods of process instability, *statistical product control* is being applied.

Tolerance: The amount of allowable variation about the nominal. *Tolerances* are often imposed without any quantitative knowledge of the capability of the process.

Upper control limit: The calculated upper boundary of a control chart representing the plus three sigma limit of the normal distribution.

Variable data: Data measurable on a continuous and incremental scale. The length, the diameter of a wire, and the elongation of a material are all examples of *variable data.*

Variation: Measurable deviation about a known point.

X Bar chart: The X Bar (Averages) chart records the averages of groups of product selected over time. When the pattern of the averages is compared to certain criteria, a determination may be made as to process stability. The control chart limits of the X Bar chart do not directly relate to the total amount of normal variation of the product characteristic being studied.

Index

N

O

P

T

U

V